The Physics of Computing

Luca Gammaitoni

The Physics of Computing

Springer

Luca Gammaitoni
NiPS Laboratory
Università di Perugia
Perugia, Italy

ISBN 978-3-030-87107-9 ISBN 978-3-030-87108-6 (eBook)
https://doi.org/10.1007/978-3-030-87108-6

This Springer imprint is published by the registered company Springer Nature Switzerland AG
The registered company address is: Gewerbestrasse 11, 6330 Cham, Switzerland

Preface

This book is about the **physics of computing** that is not the same as the **physics of computers**.

Indeed, there are many good books that discuss in detail the functioning of modern computers, with specific reference to the physical principles that rule their operations and this book is not one of them. Here, we are not going to go through any description of semiconductor materials or leakage currents in the last generation of field-effect transistors. You will not be bored with the description of the MOS capacitor or with the lithographic technique required to squeeze more transistors inside the same microprocessor. If you are interested in these arguments, please do not read this book.

How is the physics of **computing** different from the physics of computers? Well, it is like to say that we are going to talk about the physics of planetary motion; i.e. we are not interested in the dynamics of Jupiter or Mars, but we want to focus our attention to those principles that govern the motion of any of the planets, ignoring the differences among them. In this sense, we are interested in dealing with the fundamental underlying principles (if any) that characterize the computing machines existed, existing and, by definition of "fundamental underlying principles", that will ever exist in future.

Such an exaggerated ambition would require a clear and strong definition of what we mean with the words "computing" and "computer". Well, we will try to get away with the following definitions: "computing is what a computer does" and "computer is a device that can be instructed to carry out sequences of arithmetic or logical operations automatically" (source: Wikipedia). And we will give for granted that the reader knows what an arithmetic or logical operation is, although we will come back to this point later in the book. Based on these definitions, we can say that computing is merely carrying on a sequence of arithmetic and/or logic operations.

Thus, in this book, we will discuss the fundamental physics principles that rule ANY physical device built to carry on arithmetic and/or logic operations.

Perugia, Italy

Luca Gammaitoni

Acknowledgements

The author owes a debt of gratitude to all friends, colleagues, collaborators and students of Noise in Physical Systems (NiPS) Laboratory at the University of Perugia, for developing the vision which underlies the writing of this book. In particular, he wishes to thank Markus Buttiker, Giovanni Carlotti, Francesco Cottone, Davide Chiuchiù, Alessandro Di Michele, Cristina Diamantini, Miquel Lopez-Suarez, Igor Neri, Marco Madami, Fabio Marchesoni, Maurizio Mattarelli, Carlo Presilla, Carlo Trugenberger, Helios Vocca and Angelo Vulpiani. Special thanks to Igor Neri for a careful reading of the ms.

Contents

1 **Computing with Devices** 1
1.1 What Do We Mean with Computing? 1
1.2 Calculus Means Stone 4
1.3 On Computing Devices 6
1.4 Universal Machines 8
1.5 Interesting Questions 9
References 10

2 **The Physics of Digital Computing** 11
2.1 Digits 11
2.2 Binary Digits 13
2.3 Binary Switches 14
2.4 Logic Switches 17
2.5 Logic Gates 20
2.6 Logic Gates Made by Logic Switches 25
2.7 A Dynamical Model 28
2.8 The Switch 30
References 35

3 **What is Friction** 37
3.1 Energy Conservation 37
3.2 Brownian Motion 39
3.3 Langevin's Approach 40
3.4 Fluctuating Cantilever 44
3.5 Fluctuation and Dissipation Come Together 46
3.6 Thermodynamics for Everybody 49
3.7 Energy and Entropy at Small Scales 53
3.8 Friction at Small Scales 57
References 58

4 The Physics of Small-Scale Switches ... 59
4.1 Logic Switches as Thermodynamic Systems ... 59
4.2 A Digression: Energy Dissipation is an Illusion? ... 60
4.3 The Switch Dynamics in the Presence of Fluctuations ... 62
4.4 Combinational Switches: The Switch in the Presence of Fluctuations ... 65
4.5 Sequential Switches: The Switch in the Presence of Fluctuations ... 66
References ... 68

5 Fundamental Energy Limits in Computation ... 69
5.1 How Much Energy Does It Take to Switch and Why This is Important? ... 69
5.2 Combinational Switches: Minimum Dissipated Energy ... 70
5.3 Sequential Switches: Minimum Dissipated Energy ... 73
5.4 Bit Reset: Minimum Dissipated Energy ... 75
5.5 Minimum Dissipated Energy: How Far Are We? ... 78
References ... 80

6 Storing Data ... 83
6.1 Memories: How Do They Work? ... 83
6.2 Memory Dynamics ... 85
6.3 Fundamental Energy Limits for Memory Storage ... 86
6.4 Fundamental Limits for Memory Preservation ... 90
6.5 Beyond Digital Memories ... 91
References ... 92

7 What About Information? ... 93
7.1 The Big Picture ... 93
7.2 Information: What is It? ... 94
7.3 Landauer's Principle: Long Story Short ... 95
7.4 Sub-k_BT Microelectromechanical Irreversible OR Logic Gate ... 97
References ... 101

8 Cloud Computing and Data Transmission ... 103
8.1 A New Name for an Old Situation ... 103
8.2 Signal Transmission ... 104
8.3 Minimum Energy for Transmitting One Bit ... 106
Reference ... 108

9 Analog Computing ... 109
9.1 What is Analog Computing? ... 109
9.2 Analog Computers Came First ... 111
9.3 Electronic Analog Computers ... 113

9.4 Energy Dissipation and Entropy Reduction in Analog Computing ... 116
References ... 117

10 Computing in the Quantum Regime ... 119
10.1 How is Quantum Computing Different from Classical Computing ... 119
10.2 Quantum Computation and Energy Dissipation ... 121
10.3 Quantum Binary Switches ... 123
10.4 Quantum Limit in Computing Speed ... 124
10.5 Quantum Limit to Entropy Reduction in Analog Computing Systems ... 125
References ... 127

11 The Future of Computing and Society ... 129
11.1 Reflections on the Computing Era ... 129
11.2 From the Introduction of Digital Technology to Big Data ... 130
11.3 Big Data and AI for the Creation of New Knowledge ... 132
11.4 The Surveillance Society Nightmare ... 134
11.5 Are Digital Data Inevitable? ... 135
11.6 Computers of the Future: The Atoms Strike Back ... 136
References ... 138

Chapter 1
Computing with Devices

It is unworthy of excellent men to lose hours like slaves in the labor of calculation which could safely be relegated to anyone else if machines were used.
Gottfried Wilhelm Leibnitz (1646–1716)

Abstract Computing is a natural act for most of us, nowadays. However, it has not been so for many years, for many people, along the history of humankind. In this chapter, we discuss what do we mean with *computing* and how this activity can be done with the help of devices. The role of this chapter is to lay the foundations for the development of a treatment of computational processes in terms of the functioning of the physical systems that these processes are called upon to carry out.

1.1 What Do We Mean with Computing?

The term *computing* comes from the Latin *computare*, a word composed by the union of *cum* (with) and *putare* (to evaluate in quantitative sense). Together they mean to "evaluate quantitatively things", strict synonymous of *calculating*, another Latin originated word that we will consider later on. So, based on this understanding, we could say that computing something has to do with dealing with the quantitative aspects of things.

How do we do that? Since when we were kids, we had to deal with numbers. A number represents a feature of specific properties of ensembles of things. For example: a bunch of cookies in a basket comes with some properties. They are sweet, they may be warm or cold, they may be light or heavy and they may few or many. These properties are evaluated here qualitatively but, if we want to be more precise, we could specify how much these properties are in place. Given our basket of cookies, we can say that they are warm at the touch (more precisely their temperature is 42 °C), light if we raise them (more precisely their weight is 320 g), they are many (more precisely they are ten). How do we establish these quantities?

This can be done once we set a unit of measure and a method to compare the same property in different baskets. *Comparing* is clearly the first attempt to deal

L. Gammaitoni, *The Physics of Computing*,
https://doi.org/10.1007/978-3-030-87108-6_1

with quantities and *counting* is a very basic comparison. Let us look at how we determine how many cookies are in the basket by performing the counting. We start with the first cookie that comes to hand, and we set our counter to one, and then we increase the counter by one unit for every other cookie in the basket, while marking or displacing the cookies that have already been counted, in order to avoid considering the same cookie more than once. For the counter, we could use our memory or, if we have too many cookies or a lousy memory, we could use some device. The first device that comes handy is… our fingers. Things like cookies can have their *quantity* easily established with the help of fingers. Kids fingers represent a useful device for cookies, candies, sheep and other small ensembles. When the ensemble is made by many units, then it is wise to use other, more suitable, devices.

One simple solution, instead of using fingers, is to use small stones. Small stones can come in large quantities, are available for free and can be easily carried around. They can be employed very much like fingers, by associating a stone to each unit that we want to count. By accumulating stones into a basket, we can promptly count cookies, sheep and all sort of ensembles.

Even more effective than the small stones (loose or organized in an abacus) is the device represented by *writing*. This can be done by attributing a written symbol to a given quantity. We can start with a vertical mark for a unit. The repetition of this symbol is the equivalent of adding another finger. For the sake of convenience, we could use other symbols for larger quantities, like V for five units or X for ten units. As you can imagine, in this way, you can more easily take count of larger and larger quantities. This writing activity, aimed at facilitating the counting task, is also useful for keeping memory of the previously counted quantities. In short, we have introduced a way to represent quantities by associating to a given number a given symbol. A task that, in the history of mankind, has been assigned a name (numeration system) and it has been performed in different ways.

By looking at the general features of numeration systems, we can identify two main classes: *positional* and *non-positional* systems. A good example of non-positional system is the Roman numeration. Ancient romans used to represent quantities up to one thousand by using only a quite limited number of different symbols: I (one), V (five), X (ten), L (fifty), C (one hundred), D (five hundred) and M (one thousand). The interpretation of a written quantity is performed on the base of few simple rules:

(1) Start from the leftmost symbol.
(2) If this symbol is repeated, add the values.
(3) If the symbol: I precedes the V then the quantity is four, I and X (nine), X and L (forty), X and C (ninety), C and D (four hundred), C and M (nine hundred)
(4) Add all the obtained values.

As an example: XXIX represents twenty-nine (XX = twenty plus IX nine), CCCXLVII represents three hundred forty-seven.

On the other hand, a good example of positional system is represented by the decimal numeration, who uses ten different symbols: 0 (zero or no quantity at all), 1 (one), 2 (two), 3 (three), 4, 5, 6, 7, 8, 9, whose value is obtained by considering the position of each symbol in the written representation. As you are familiar: 136

means one hundred thirty-six. This is obtained by adding quantities associated at each symbol according to the following rules:

(1) Compute the value of each symbol. The value is represented by the value of the symbol, times the base (the number of different symbols, in this case ten) multiplied by itself many times as the number of symbols at its right minus one. When there is no other symbol, the value to be multiplied is one).
(2) Add all the obtained values.

As an example, 1342 represents one thousand (one times ten multiplied by itself two times) plus three hundred (three times ten multiplied by itself one time) plus forty (four times ten multiplied by itself zero times) plus two (two times one).

The same system can be applied to a base different from ten. A popular example is represented by base two. In this base, we have only two symbols "0" and "1" with the same value as in base ten. In this base, the string 1001 represents the quantity nine, and it is obtained, according the same rules as before: starting from the leftmost symbol, we have eight (one times two multiplied by itself two times) plus zero (zero times two multiplied by itself one time) plus zero (zero multiplied by itself zero times) plus one (one times one).

The positional numeration system has some advantages that will appear clear in the following of the chapter and thus has eventually become the most adopted system in modern times. One important aspect to be noted is that in the positional system, we have a symbol that represents the null quantity that was not present in the Roman system.

Back to our *counting* task, there is no doubt that the device represented by writing is of great help in carrying out this task. Once we have done the counting and, with the help of the writing, we have represented the quantity that belongs to a given ensemble, now we are faced with the computing task itself. How do we deal with these quantities?

Common ways of dealing with quantities include adding, i.e. putting together different baskets or separating them by dividing a certain quantity in a given number of baskets, for example. All this can be done by using special procedures that we call *algorithms*.

The word *algorithm* comes from Arabic and refers to the name of the ninth-century mathematician Muḥammad ibn Mūsā al-Khwārizmī, subsequently Latinized as *algoritmi*. These procedures have been in place since very long time and were used already by ancient Babylonian mathematicians c. 2500 BC and Egyptian mathematicians c. 1550 BC.

1.2 Calculus Means Stone

To give an example of algorithm, we could consider the following procedure. Let us suppose that we have a flock of sheep. We count them by using stones, and we find that they are six hundred twenty-five. Now, our neighbour sells us his own flock that is made by two hundred eighty-one sheep. How many sheep do we have altogether?

Let us see a possible algorithm using the small stones device. This choice is historically relevant because one of the first device employed was actually made by small stones. The word "calculus" itself is the Latin word for small stone (Fig. 1.1).

These are the required steps:

We put one stone, for each sheep in our flock, into a basket.

We put one stone, for each sheep in the acquired flock, into the same basket.

Finally, we count how many small stones there are in the basket.

It is a simple process, indeed. And most importantly, it is a process that, if followed step by step by any person, can bring to the same, correct solution.

Thus, we can say that an *algorithm is a procedure that should be followed in proper order to reach the desired goal.*

If we change the device, also the algorithm changes. As an example, we discuss a possible algorithm for the device "writing". We use the base ten notation and the algorithm is the following:

We count the sheep in our flock. They amount to 625 sheep.

We count the sheep in our neighbour. They amount to 281 sheep.

We write the two representations in column by aligning towards right:

Fig. 1.1 Small stones represent one of the first and simplest computing devices. (Photo by Wolfgang Hasselmann)

625

281

We add the rightmost two symbols of each number. If the result is smaller than ten, we right below in the same column the symbol equivalent to the result. If the result is not smaller than ten, then we right the rightmost symbol of the result and carry-on the other symbol for the next column.

625

281

6

We now proceed to add the two-minus-one symbols of each number with the carry-on (if any). If the result is smaller than ten, we right below in the same column the symbol equivalent to the result. If the result is not smaller than ten, then we right the rightmost symbol of the result and carry-on the other symbol for the next column.

625

281

06

We continue the same procedure until we reach the last digit of the largest number.

625

281

906

The final result can be read in the new number that has been assembled with this process: 906 (nine hundred and six).

It is interesting to note that this algorithm works in the same way, regardless the base that we are using. As an example, we look at the addition between the two quantities in base two: seventeen (10001) plus thirty-nine (100111).

10001

100111

111000

The result is fifty-six (111000), and it is, clearly, the same result if we apply the same algorithm to seventeen (17) and thirty-nine (31) that gives fifty-six (56).

Thus, we have seen that computing, that means dealing with quantities, is made through algorithms, and these algorithms depend on the device that we use to do computation.

1.3 On Computing Devices

So far, we have seen that loose stones represent a computing device, and also, writing is a device. There exist other devices that have been invented along history in order to make the computing activity as easy as possible and widely available to many people.

In the available literature, both on the Internet [1] and in books [2], there is plenty of information about the history of computing devices; thus, we will not present here a detailed account for it. Instead, we will focus our attention on a simple device, whose behaviour will be useful in the further development of the book: the *slide rule*.

The modern slide rule is a mechanical computing device that descends from a long tradition of devices based on sticks with marks. The most ancient predecessor is probably a member of the family of so-called *tally sticks*.

These were sticks made with bones that presented horizontal marks aimed at recording quantities. Among them, the *Ishango bone* is probably one of the most famous. It is an animal bone approximately dated around 20,000 BC. Originally found in Congo, in the sixties, it is made with the fibula of a baboon and has a sequence of tally marks carved in three rows. Another example, even older, is the *Lebombo Bone*, dated around 40,000 BC is a baboon's fibula with 29 distinct notches, found in a cave in the Lebombo Mountains of Swaziland.

Modern slide rules are usually made by three parts (see Fig. 1.2):

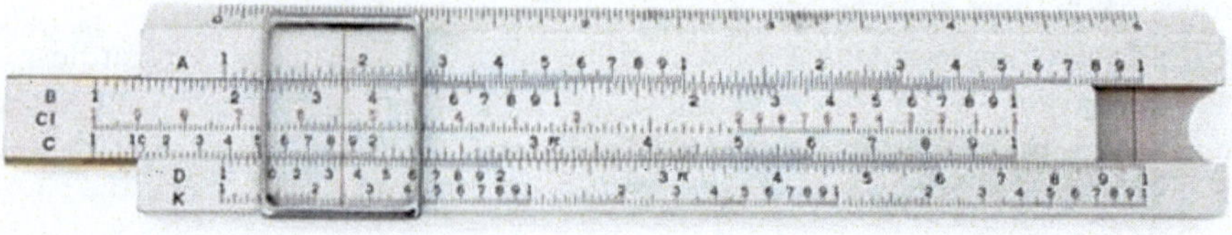

Fig. 1.2 Slide rule

- The base that consists in two parallel sticks, connected together and separated in between by a channel that hosts a slide.
- The slide, a third stick that can move along its axis, in parallel with the base.
- The runner, an exterior sliding, transparent window with a vertical line.

The functioning of the slide rule is quite simple. Let us suppose we want to multiply two numbers, say 2 and 4. What we can do is to use two sticks that are marked along their length with marks representing the logarithm of the integer numbers, like in Fig. 1.4.

We can displace the two sticks in order to add their lengths and look for the total length. In Fig. 1.4, we shift the upper stick (orange) in order to set its start at the number 2 and then we add a length corresponding to the number 4. In correspondence with 4, in the lower stick (cyan), we read the result of the multiplication that is 8. This is made possible by a well-known property of the logarithmic function:

$$\log(2) + \log(4) = \log(2 \cdot 4)$$

In this way, we can perform multiplication by adding lengths.

Fig. 1.3 Top, grey: Lebombo bone. Bottom, brown: Ishango bone (J.D. Loreto and D.H. Hurlbert Smithsonian)

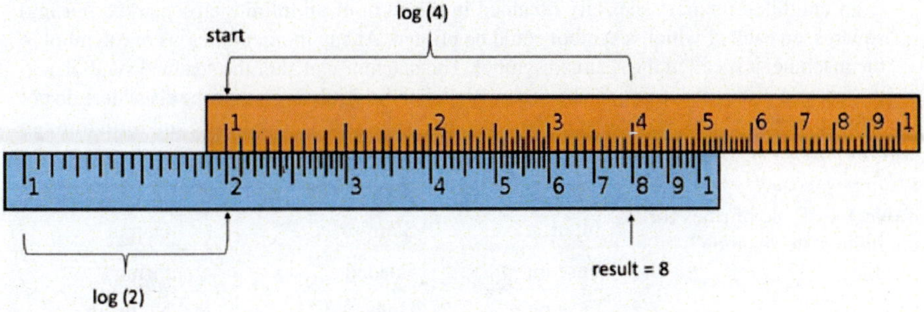

Fig. 1.4 Functioning of the slide rule: multiplication by adding lengths

This is just an example, and more complicated operations can be performed with the slide rule, like division, trigonometric functions, logarithms and exponentials.

Once again, we have a computing device and a set of associated algorithms, all new and different compared to the previous devices that we have mentioned.

1.4 Universal Machines

Clearly, along the history of man, many different computing devices have been invented and realized. The modern digital computer is only the most recent one in a long list of devices.

Some of these devices have been conceived for special purposes. A good example is the Turing machine. An ideal device conceived by Alan Turing in 1936 and published in the Proceedings of the London Mathematical Society in early 1937. Technically speaking, a Turing machine is an ideal finite-state machine that can interact with an infinite length tape, where are written symbols. One symbol at a time can be read and, subsequently written by the machine, according a table of rules that considers the read symbol and the machine internal states.

To get a better grasp of the special device called Turing machine, let us start with what is the *state* of a machine. Let us think at the functioning of an old-style big gun. In the old times, a cannon gun could be in three different *states*: loaded and ready to fire, unloaded and out of service.

We can imagine that our device has an input (the cannon fuse) and an output (the firing of the cannon ball). If we set fire to the fuse, the output will be the firing of the ball or nothing at all, depending on the state of our machine, according to the following table of rules (Table 1.1).

Most importantly, the *state* of the machine can change with time as a consequence of the operations. In our example, the cannon state changes from loaded to unloaded after a firing event.

In a similar way, the Turing machine has states that can be changed during computation. In the very same words of Alan Turing, its machine requires:

> …an unlimited memory capacity obtained in the form of an infinite tape marked out into squares, on each of which a symbol could be printed. At any moment there is one symbol in the machine; it is called the scanned symbol. The machine can alter the scanned symbol, and its behavior is in part determined by that symbol, but the symbols on the tape elsewhere do not affect the behavior of the machine. However, the tape can be moved back and forth through

Table 1.1 Table of rules for the finite-state "cannon" machine

Input	State	Output
Fuse on	Loaded	Firing
Fuse on	Unloaded	No firing
Fuse on	Out of service	No firing
Fuse off	Any	No firing

> the machine, this being one of the elementary operations of the machine. Any symbol on the tape may therefore eventually have an innings [3].

To be more precise, we can say that the ideal Turing machine is composed by the following parts:

- *Tape* of infinite length and divided into cells each containing a symbol taken from a finite set of symbols.
- Reading/writing *head* that can read/write symbols on the tape.
- *Moving mechanism* to shift the tape left or right, one step at the time.

At every step, a symbol is read, and the machine is operated according a table of rules. As an example: *if the symbol is* α *and the state is A, then shift the tape right*. Or *if the symbol is* α *and the state is B, then erase* α *and write* β*, change the state from B to A and shift the tape left*, and so on.

The important aspect about this quite peculiar device is that Turing demonstrated that with this machine, any computable algorithm can in fact be performed. In this sense, the Turing machine represents a universal computing device. At difference with the other devices that we have seen before, the Turing machine can compute any algorithm and thus represents a very useful instrument for inspecting the limits in what can or cannot be computed.

1.5 Interesting Questions

In the previous paragraphs, we have discussed what do we mean with computing and how this can be done with the help of devices. As we have seen, an important ingredient is represented by the algorithms that is the set of rules to be followed in order to perform a given computation.

Clearly at this time, by the point of view of physics, we can ask all kind of interesting questions about the act of computing. Questions like how long does it take to do a certain computation or how much work do we need to perform in order to compute a given operation and, most importantly, how do these answers depend on the device that we consider or the algorithm that we use.

Some of these questions have been around for many years and the many attempts to provide meaningful answers is an important part of what is today considered *computer science*.

In the following of this book, we will take on some of these questions with the specific goal to find answers that are, as much as possible, unrelated to a given algorithm or to a given device.

As an example, we are going to address the following question:

> What is the minimum energy required to perform a computation?

In dealing with this question we will try to find an answer by specifically considering the "minimum" condition. More precisely it is our intention to show that, according to the fundamental principles of physics, the minimum required lost work to be performed or, what is the same, the minimum energy to be dissipated, in performing a computation, is zero.

By the moment that this minimum is zero and it is also the least that you can do when you operate any physical system, due to the principle of conservation of energy, we do not need to show that this is true for any physical device. All we need to show is that this is possible for one given device and that this device can be operated as a universal machine, i.e. capable of processing all computable algorithms. If this is true, clearly, any other device will be capable of the same or it will be worst. In any case we will have reached our goal.

We will reach this result by analysing the way present digital computer works and reducing all the performed operations to a sequence of elementary operations whose functioning requires zero energy. This reduction to elementary operation will be performed by addressing the functioning of universal logic gates and showing how this functioning is the result of the combination of the operation of more basic components called binary switches. But let's not anticipate too much what we are going to show in the next few chapters: keep reading and you will see.

References

1. See e.g. Wikipedia at https://en.wikipedia.org/wiki/History_of_computing_hardware
2. P.E. Ceruzzi, *Computing: A Concise History* (The MIT Press Essential Knowledge series, 2012)
3. A.M. Turing (Jul 1948). Intelligent Machinery (Report). National Physical Laboratory, pp. 3–4

Chapter 2
The Physics of Digital Computing

Computer Science is no more about computers than astronomy is about telescopes.
Edsger W. Dijkstra

Abstract There is no doubt that *quantity* is one of the most important features of the world, as we know it today. According to what we have seen in the previous chapter, computing is an activity that involves dealing with *quantity*. Specifically, we aim to evaluate quantitatively things around us. A major step in this direction was taken thanks to the introduction of an efficient representation for these quantities. In this chapter, we will discuss digits, i.e. numbers and their representation. Most importantly, we will introduce a conceptual device, the *binary switch*, that is the very basic brick of any computing architecture. Once we have designed a device, we can think of a physical system that makes it and, therefore, of the laws of physics that regulate its functioning.

2.1 Digits

In this chapter, we will deal with digital computing. But, what *digital* does it really mean? Well, the word digital comes (again) from the Latin *digitus* meaning finger. In this respect, digital is something that has to do with the way we count with fingers. In present-day English, according to the Merriam-Webster dictionary, *digital* stays for: "of, relating to or using calculation by numerical methods or by discrete units". Such discrete units are also called *digits*. Thus, digital computing is a form of computing done with quantities represented by discrete units, like the quantities that we have discussed in the previous chapter and that we used to represent sheep or cookies. As we said, the quantities of interest could be represented by a finite number of symbols, like , in the positional system, base two, by a sequence of symbols "0" and "1".

It is a common experience, however, that not all quantities can be represented by a set of digits, like sheep or cookies. As an example, we could consider the weight of a basket of apples. If we use a balance, we can establish that their weight is 2.36 kg.

L. Gammaitoni, *The Physics of Computing*,
https://doi.org/10.1007/978-3-030-87108-6_2

However, in this case we should not be mistaken because the presence of the decimal point does not change the nature of the number representing the quantity. In fact, if we change the unit of measure from kilogram to gram, we can write that the weight of the apples in the basket is 2360 g. Once again, the quantity is expressed by a set of digits, as before.

However, at a certain point in history, we discovered that (unfortunately) there exists quantities that resist this kind of approach. One of the most famous examples is $\sqrt{2}$. Around the fifth century BC, the problem at hand among the Pythagoreans was to find a common unit to measure both the side and the diagonal of a square, i.e. to represent with a finite number of digits the length of a diagonal of a square of unitary side. The discovery that this was not possible is attributed to Hippasus of Metapontum and literally produced great scandal inside the Greek school. The legend tells that he realized the existence of the irrational numbers while he was at sea with his colleagues. The discovery was against the Pythagorean doctrine that claimed that all the quantities of the universe had to be expressed by rational numbers (i.e. numbers that can be represented with a finite number of digits). As a result, the discovery was unbearable and they decided to throw Hippasus overboard.

Another popular irrational quantity is the ratio between the circumference and the diameter of a circle: the famous π. The irrational number π has also another interesting property: it is a *transcendental*, rather than an *algebraic*, irrational number. This implies that it cannot be represented as a solution of an algebraic equation with rational coefficients. The story of this discovery is associated with a very famous problem called *squaring the circle*. The problem was initially identified by the ancient Greeks and can be stated as follows:

> given a circle, draw the square that has the same area of the circle.

Although the Greeks were the first to clearly see a problem in this task, earlier attempts were put in place by the Babylonians and by the Egyptians that proposed approximate (and wrong) solutions. In more mathematical terms, if we indicate the area of a circle of radius r with πr^2, then we need to find the side l of a square such that $l^2 = \pi r^2$, whose solution is represented by

$$l = r\sqrt{\pi}$$

The sad news is that no matter how hard you try, also the quantity $\sqrt{\pi}$ cannot be expressed by a finite number of digits, as we have done before for the number of cookies or the weight of a basket of apples. Although the problem was known for a very long time, we had to wait until 1882 when the impossibility was actually proven by the Lindemann–Weierstrass theorem.

Based on what we said, it is now clear that digital computers can deal only with rational quantities, i.e. quantities that can be expressed by a finite number of digits.

What can we do, then, if we have to do with quantities that are not rational? We do the same procedure that we use when we have to deal with quantities obtained by an experimental measure: we approximate.

As an example, let us consider the irrational number $\sqrt{2}$. Its representation in base ten can be written as:

$$1.414\,213\,562\,373\,095\,048\,80...$$

The suspension points here indicate that we have an infinite number of successive digits. By the moment that in any digital computer we have only a limited number of digits, we need to approximate this quantity. Let us suppose that we have seven digits, then we will have 1.41421. At this point, clearly, we are introducing a possible error because in our calculation we are not using the correct quantity $\sqrt{2}$. The error that we introduce is called *representation error* and is measured by the difference between the actual quantity and its approximation. Such an error can be reduced by increasing the number of digits but, no matter how many digits we put in, there will always be a nonzero error that will eventually impact the result of our calculation by combining and propagating with other approximation errors: a phenomenon that is known as *roundoff error*.

Is there any way to avoid errors associated with the finite number of digits? The answer to this question is clearly no, unless we leave the realm of digital computing and move into analog computing. But we will have to wait until we reach Chap. 9 where we will discuss this alternate form of computing.

2.2 Binary Digits

We have already introduced the base two numeration system. A quantity written in this way is usually expressed as a sequence of only two symbols: "0" and "1". For this reason, we call them bits, a contraction of "binary digits". Present time digital computers use a representation of quantities based on bits. There is more than one reason why base two is, in general, preferable to other bases. One of the most relevant is associated with the physical realization of the digits. We will postpone a more detailed discussion of this point until later. For the moment, let us simply assume that it is easier to make a computer work if we use just two different symbols for representing quantities.

Let us spend few more words on how we can use the base two numeration system, to represent quantities to be used in digital computers. First of all, we should keep in mind that, as we have said above, any digital computer uses a finite number of bits. Old time digital computers used a representation based on only groups of 8 bits (named *byte*). With 8 bits, we can represent all the numbers between 00000000 and 11111111 and it is quite easy to check that these are all the numbers that go from 0 to 255, in total 256 different quantities. If we change the number of used bits to n, it changes this number according to the simple rule

$$\text{Number of different quantities} = 2^n.$$

If we use these bits to represent positive numbers, then we will have all numbers from 0 to $2^n - 1$.

How about negative numbers? In this representation, we do not have a symbol for signalling that the quantity that we are representing is negative; thus, we invented a special representation called *two's complement*.

Two's complement representation of a given number m is obtained according to the following rules:

- if m is positive, usual representation in bits
- if m is negative, usual representation in bits of the number $2^n - m$

Let us see some examples with 8 bits:

+5 is represented as 00000101.

−5 is represented as 256−5 = 251 and thus 11111011.

With n bits in the two's complement representation, we have still 2^n different quantities and precisely all the numbers that run from -2^{n-1} to $2^{n-1}-1$.

The great advantage of this method is that addition and subtraction can be performed at once with the same algorithm that we have already discussed. Let us see some examples with 8 bits:

5−3 = 00000101 − 00000011 = 00000101 + 11111101 = 00000010 that corresponds to 2.

3−3 = 00000011 00000011 – 00000011 + 11111,101 = 00000000, right as expected.

Finally, we notice that to find the two's complement at m, we could also use a fast and dirty trick: we take m expressed in bits and change all the "0" into "1" and all the "1" into "0" and then add 1.

As an example, to find −3 we can start with +3 = 00000011. We change as described and obtain 11111100. Now, we add 1 to get 11111101. And this is the expected number.

2.3 Binary Switches

In Chap. 1, we have seen that we could use stones as a device to perform computation. In order to perform digital computation in base two, we can keep using stones to represent bits and do computation. As an example, we can consider the device illustrated in Fig. 2.1. It is a device made by a pebble and two bowls with the pebble that can be placed inside one of the two nearby bowls. We can associate bit "0" with the pebble in the left bowl and bit "1" with the pebble in the right bowl. Thus, when we perform a computation, the result will be apparent by the position of the pebble. Clearly, as the result of a computation, changing one bit here corresponds simply to translating the pebble position.

Fig. 2.1 A pebble and two bowls are all that it takes to represent one-bit device. Bit "0" is associated with the pebble in the left bowl (upper picture), and bit "1" is associated with the pebble in the right bowl (lower picture)

This simple device, indeed, encompasses the very basic ingredients that are required to perform any possible digital computation, as it will be soon clear. By generalizing the functioning of the *pebble–bowls device*, we can say that in order to perform digital computing we need two fundamental ingredients:

1. a *physical system* capable of assuming two different *physical states*
2. a way to induce state changes in this physical system (typically a force)

With *physical system,* we intend here any object, device or phenomena that can be studied by physics. A *physical state*, on the other hand, is a given set of measurable quantities, whose value can be used to distinguish unambiguously two different outcomes. In our *pebble–bowls device,* we have:

1. The *physical system* is the pebble with two bowls, and the two *physical states* are represented by vales assumed by the measurable quantity "position of the pebble":
   - state "0" = pebble in left bowl
   - state "1" = pebble in right bowl
2. The way to induce state changes is represented by a force that moves the pebble from one bowl to the other.

According to this example, we can perform digital computing simply by changing the position of the pebble, according to certain rules, with the underlining idea that, while we do these changes, we are at the same time changing the value of the symbol "0" and "1" associated with the system state. Devices that obey the rules (a) and (b) are called *binary switches.*

In a more general and abstract way, we can consider any physical system (blue box in Fig. 2.2) that can assume a numerable set of different physical states. These

Fig. 2.2 A physical system that can be in different physical states

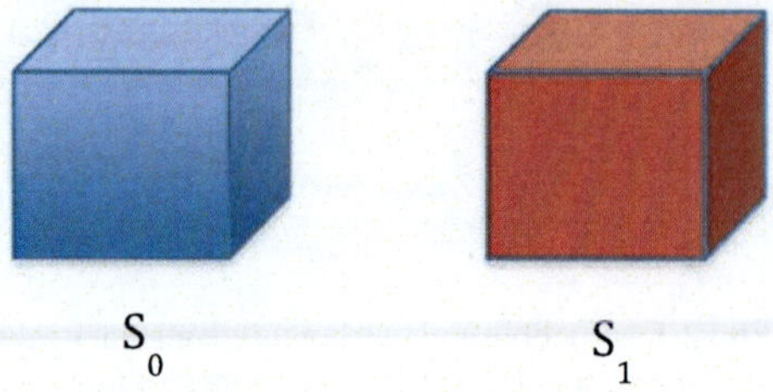

Fig. 2.3 Two physical states are illustrated here by different colours of the same box

can be identifiable as a different set of physical properties as electric charge, voltage, temperature, pressure, magnetic momentum, etc.

When we deal with binary digital computing, we need two distinguishable states, like in Fig. 2.3.

As we have anticipated above, digital computation is associated with system transformation. According to physics, in order to change the state of a physical system we need to apply a *generalized force* to it. As an example, if the state is associated with electrical voltage, then we need to apply an electromotive force that displays some charge and, as a result, the voltage of the system is changed.

The force that is something that clearly comes from outside our physical system, in computer science language, is often called *input*. In this language, we provide some input to our binary switch and, as a result of this input, a given state of the switch is produced. As we illustrate in Fig. 2.4, with an example, a given input can produce two distinct results: either the state is changed or not.

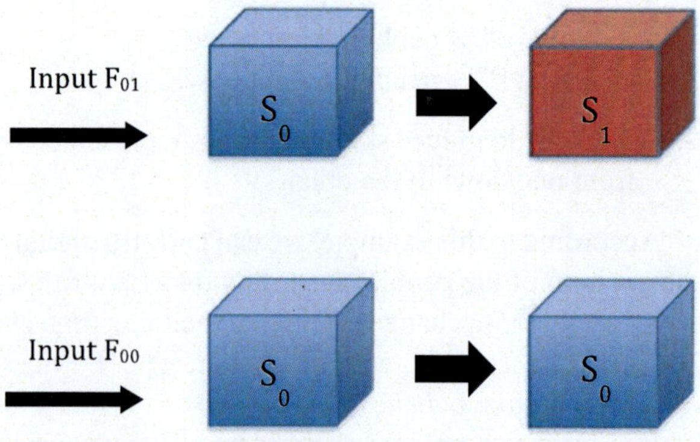

Fig. 2.4 A given input, depending on its value, can produce two distinct results: either the state is changed or not. The result state of a given input is often indicated as an output produced by that input

In computer science language terms, this amounts to say that a given input produces a given output. Input–output relations are defined in order to perform the desired calculations, as it will be clear in the next paragraph.

2.4 Logic Switches

Our *pebble–bowls device* presented in the previous paragraph clearly represents a possible realization of a binary switch. We associate bit "0" to the physical state S_0 and bit "1" to the physical state S_1. Sometimes, these are called *logic states* and the physical device that realizes such states is called *logic switch*.

According to the principle of operation of logic switches, we can identify two distinct families of switches: *combinational* and *sequential* switches. These names are usually attributed to networked logic gates (see next paragraph), but in this book we will use them to identify two families of logic switches.

Combinational switch: In the absence of any external force, under equilibrium conditions, the switch is in the state S_0. When an external force that we will call F_{01} is applied, it does switch to the state S_1 and remain in that state as long as the force is present. Once the force is removed, it does go back to the state S_0.

Sequential switch: In this case, in the absence of any external force, under equilibrium conditions, the switch can be in state S_0 or in state S_1. The logic state of the switch can be changed from S_0 to S_1 (S_1 to S_0) by applying an external force F_{01} (F_{10}). Once it is in the state S_1 (S_0), it remains in this state even when the force is removed.

As an example of *combinational* switch, we may consider a mechanical cantilever (see Fig. 2.5).

Here, the position of the tip of a mechanical blade, fixed at the other extreme, can be employed to encode a logic state ("0" or "1") into a physical state: S_0 (no bending) or S_1 (some bending). By setting a bending threshold, in order to distinguish the two

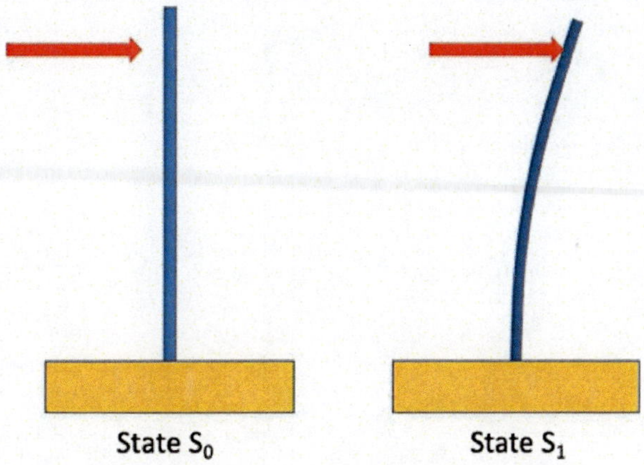

Fig. 2.5 A mechanical cantilever (blue) can be displaced from its initial vertical position (state S_0) into a bended position (state S_1) by an external pushing force (red arrow)

logic states, we can easily realize a combinational switch: once the force is removed, the switch goes back to its initial state thanks to the elasticity of the cantilever.

Note that the association of a logic state to a physical state is completely arbitrary and purely conventional. In fact, we might have reversed the states S_0 and S_1 by associating the non-bended state to the logic "1" and the bended state to the logic "0". We can also change the way the cantilever is operated. As an example, let us consider the configuration in Fig. 2.6.

Another good example of *combinational* switch is represented by the electronic transistor (see Fig. 2.7). For those who are not familiar with this electronic device, we briefly mention that it was invented in 1947 at Bell Laboratories in Murray Hill, USA, by three physicists: William Shockley, John Bardeen and Walter Brattain who got the Nobel Prize for such an invention in 1956. The transistor is an electronic device whose functioning is based on the physics of semiconductor materials. It has

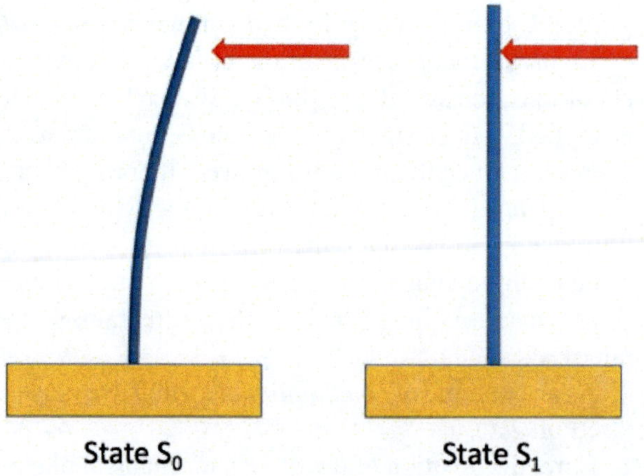

Fig. 2.6 A different configuration of the mechanical cantilever. The initial position (bended) is associated with the state S_0 and can be changed into the vertical position (state S_1) by applying an external pushing force (red arrow)

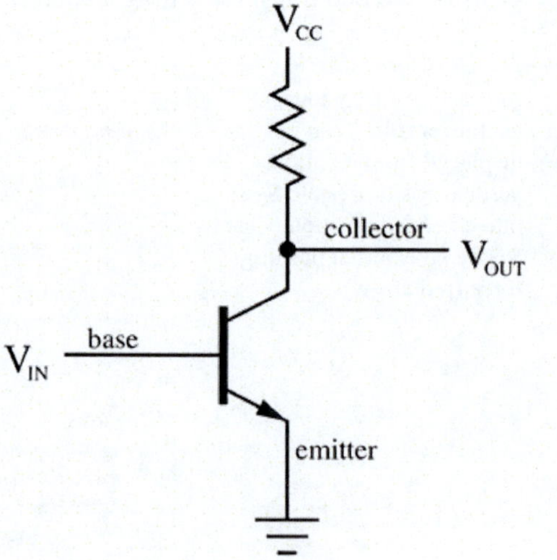

Fig. 2.7 Transistor. It can be employed as combinational switch

Fig. 2.8 Mechanical bistable toggle switch employed as a sequential switch

three distinct electrodes denominated *collector*, *base* and *emitter* (see Fig. 2.7). In very simple terms, the transistor can be operated as a binary switch as follows: when the voltage at the base is below a certain threshold V_{th}, the *collector* and *emitter* are isolated and no current can travel from V_{CC} to the ground. As a result, the voltage V_{OUT} is the same of V_{CC}. If instead, we apply a voltage $V > V_{th}$ to the *base*, the electric connection between the *collector* and the *emitter* is established and the current flows towards the ground. As a result, the voltage V_{OUT} changes to the ground voltage, conventionally assumed $V = 0$. Thus, by a applying an electromotive force to the *base* electrode we can control the value of the voltage V_{OUT} at the *collector* terminal.

If we associate the logic state "0" to a voltage $V < V_{th}$ and the logic state "1" to a voltage $V > V_{th}$, we have realized a combinational switch whose output is changed only when an electromotive force is applied at the input terminal. As soon as the force is removed, the switch goes back to its initial state.

In the family of *sequential* switches, we can certainly list the traditional mechanical bistable switch (see Fig. 2.8). Here, we can easily identify two natural states: logic state "0" associated with the "ON" position (vertical toggle to the left) and logic state "1" associated with the "OFF" position (vertical toggle to the right) or vice versa. As it is typical of the *sequential* switches, the logic state can be changed from S_0 = "0" to S_1 = "1" by applying an external force F_{01}, whose effect is to tilt the toggle from left to right. Once our switch is in the state S_1 (toggle in the right position), it remains in this state even when the force is removed. It does switch from S_1 to S_0 by applying a different force called F_{10} that brings the toggle from right to left. Once it is in S_0, it remains in this state even when the force is removed.

Combinational and sequential switches, as we will see in the following of the book, are widely used to perform all the basic operation of present-day digital computers. It is interesting to note that, although there are, in principle, many physical systems that can be employed to make good logic switches, nowadays digital computers employ almost universally electronic transistors. As we have seen above, a transistor is a good example of *combinational* switch. However, with the addition of a small capacitor we can easily make of it also a *sequential* switch. In Fig. 2.9, we present an electronic *sequential* switch.

The physical state in this case is represented by the value of the voltage at the *emitter* electrode (V_{OUT}). Let us suppose that the capacitor has initially no charge. When the voltage at the *base* is below threshold, there is no conduction from the *collector* to the *emitter* and the voltage at the *emitter* is low (state S_0). When the

Fig. 2.9 Transistor with the addition of a capacitor between the emitter and the ground. It can be employed as sequential switch

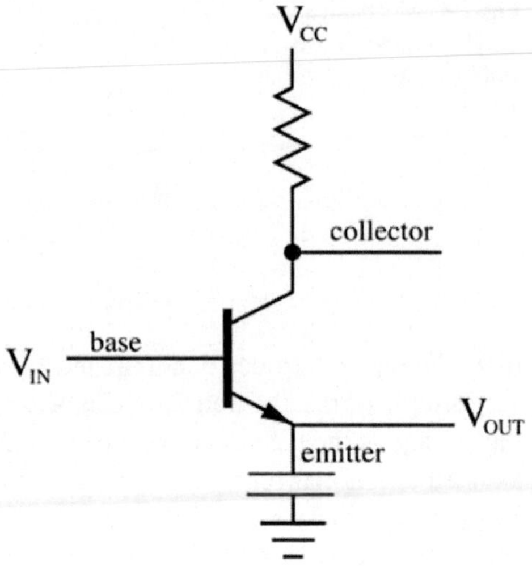

voltage at the base is above threshold, then the conduction channel is open and the capacitor is charged at the voltage V_{CC}. At this point, the voltage at the *emitter* is V_{CC} (state S_1). When the voltage at the *base* is now brought back below threshold, the voltage at the *emitter* is held by the capacitor and, thus, the state S_1 holds. When we want to go back to the state S_0, we need to change the voltage of the *collector* from V_{CC} to ground and then bring the voltage of the *base* above threshold. Now, the capacitor discharges through the *collector* and the state S_0 holds also when the *base* is brought below threshold.

Sequential switches are mainly used to represent bits, and their presence inside a computer is mostly associated with the need to store the value of a given bit and keep it for some time. On the other hand, combinational switches are mostly used to perform calculations. How this is made is the subject of the next paragraphs.

2.5 Logic Gates

When operating with binary digits, we have available at our disposal new algorithms that add to the usual set of sum/subtraction, multiplication/division. This set of additional tools for performing computation goes nowadays under the name of Boolean algebra and was introduced, for the first time, by the British mathematician George Boole in 1847 in his book *The Mathematical Analysis of Logic* (1847), and further developed through 1854 where it was described in the book *An Investigation of the Laws of Thought*. In Boole's intention, his work was aimed at establishing the rules of logic by manipulating the two opposite concepts of "true" and "false" (or "0" and "1" in our notation).

In this binary mathematical framework, Boole introduced new useful algorithms (i.e. rules) like the *conjunction* (also called AND operation), the *disjunction* (OR operation) and the *negation* (NOT operation). If we have two bits, we can compute the AND operation according to the following set of rules:

$$0 \text{ AND } 0 = 0$$

$$0 \text{ AND } 1 = 0$$

$$1 \text{ AND } 0 = 0$$

$$1 \text{ AND } 1 = 1$$

For the OR operation:

$$0 \text{ OR } 0 = 0$$

$$0 \text{ OR } 1 = 1$$

$$1 \text{ OR } 0 = 1$$

$$1 \text{ OR } 1 = 1$$

And for the NOT operation: NOT $0 = 1$, NOT $1 = 0$.

AND, OR and NOT operations can be combined to perform more complicated calculations like:

$$(1 \text{ AND } (1 \text{ OR } 0) \text{ OR NOT } (0)) = (1 \text{ AND } 1) \text{ OR } 1) = 1 \text{ OR } 1 = 1$$

In order to provide a visual description of these operations, we have also invented symbols that graphically represent the sequence of operations involved (Fig. 2.10):

In this graphical representation, the Boolean operation is assimilated to an input–output process: the input bits are present at the left of the symbol, while the output bit is present at the right. The symbol itself represents the "machinery" necessary to transform the input bits into the output one. By the moment that the computation moves from left to right, the AND, OR and NOT representations can also be seen as "gates" that let in (from left) and out (towards right) the flux of the computation. For this reason, they are also called *logic gates*.

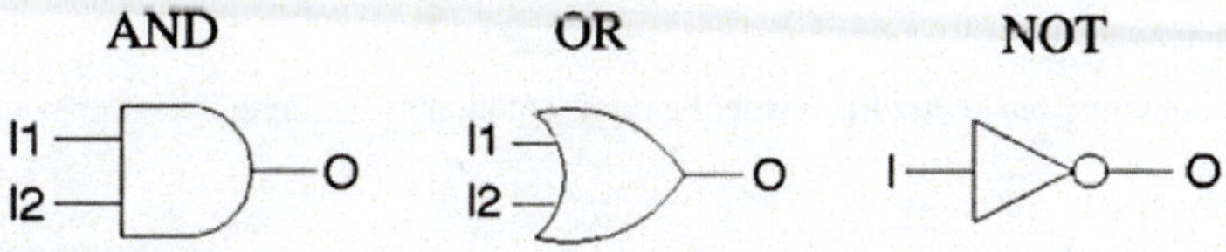

Fig. 2.10 Symbolic representation of the logic operations

The graphical representation of the logic gates provides also a pictorial representation of a more complicated calculation. As an example, the so-called *combinational network* presented in Fig. 2.11 is equivalent to the following Boolean algebraic expression:

((A AND B) OR (A AND C)) OR (C AND B)

where A, B and C are bits.

The interesting aspect of the logic gates is that we can use them also to perform the traditional algorithms. As an example, let us give a look at the one-bit addition.

According to the graphical representation that we introduced above, the addition of two bits can be represented as an input–output operation (see Fig. 2.12), where at the input we have the two addenda B1 and B2 and at the output we have the sum S and a possible carry-on R.

The addition operation can be illustrated through Table 2.1 (sometime called *truth table*) where we have explicitly considered all the possible combinations of the values of the input bits B1 and B2 and have associated in the same row the expected outcome, according to the addition operation that we discussed in 1.2.

The box in Fig. 2.12 represents the machinery of the adder. It is not surprise that such a machinery can be described as a combinational network of logic gates. In Fig. 2.13, we present such a network. The reader can verify that it obeys, indeed, to the truth table in Table 2.1.

To be precise, the combinational network in Fig. 2.13 is called "half adder". In fact, if we want to build machinery for a full adder, we need to consider also the possibility that we have a carry-on bit coming from a previous addition. To adjust

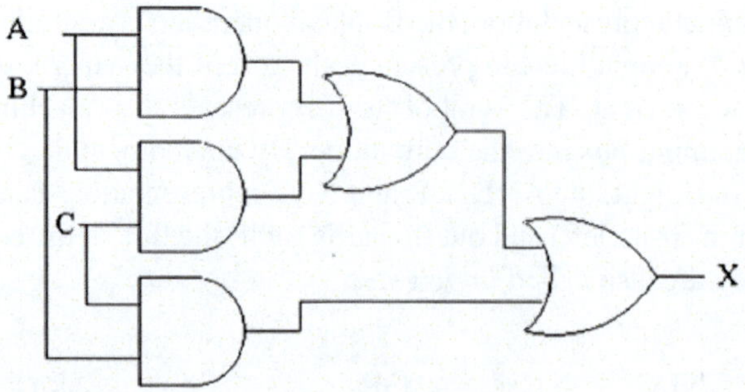

Fig. 2.11 Combinational network: Symbolic representation of the logic operations

Fig. 2.12 Graphical representation of the one-bit addition

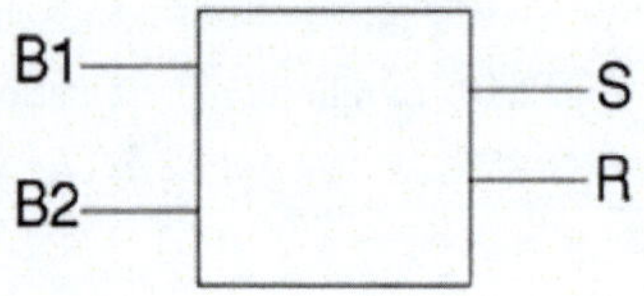

Table 2.1 Truth table for the one-bit sum

B1	B2	S	R
0	0	0	0
0	1	1	0
1	0	1	0
1	1	0	1

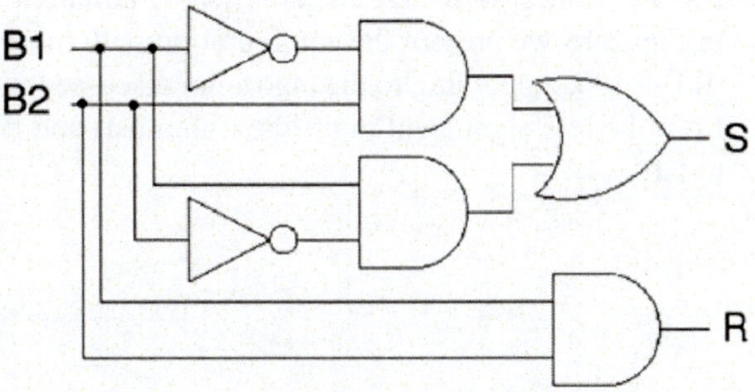

Fig. 2.13 Graphical representation of the one-bit addition

for this eventuality, we should consider a slightly more complicated truth table, as in Table 2.2, where R_{in} represents the incoming bit, i.e. a carry-on, from a previous operation.

In Fig. 2.15, we illustrate the corresponding network where we have introduced, for the sake of simplicity, a new logic gate denominated XOR or "exclusive OR" that obeys the following simple rule:

Table 2.2 Truth table for the one-bit full adder

R_{in}	B1	B2	S	R
0	0	0	0	0
0	0	1	1	0
0	1	0	1	0
0	1	1	0	1
1	0	0	1	0
1	0	1	0	1
1	1	0	0	1
1	1	1	1	1

$$0 \text{ XOR } 0 = 0$$

$$0 \text{ XOR } 1 = 1$$

$$1 \text{ XOR } 0 = 1$$

$$1 \text{ XOR } 1 = 0$$

And it is represented by the symbol in Figs. 2.14 and 2.15.

To conclude this paragraph, we illustrate a possible combination of one-bit full adders in order to realize the addition of numbers with more bits. As an example, in Fig. 2.16 we present the configuration with three bits: A1A2A3 + B1B2B3, e.g. 011 + 100. According to the algorithm discussed in Chap. 1, the computation starts from the less significant digit (the rightmost) and thus the flux proceeds from right to left.

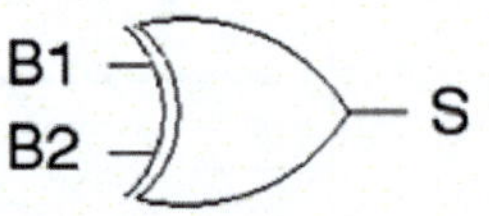

Fig. 2.14 Graphical representation of the XOR logic gate

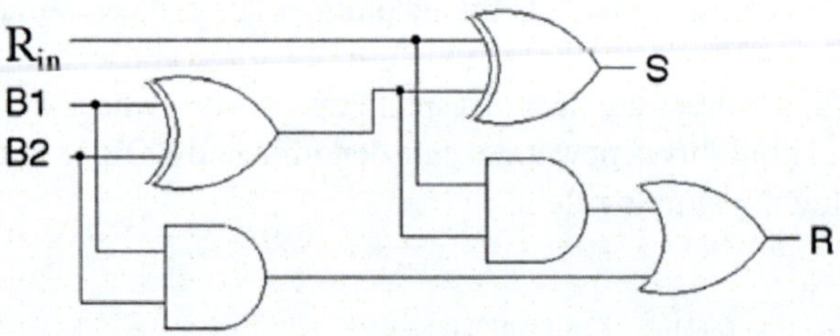

Fig. 2.15 Graphical representation of the one-bit full adder

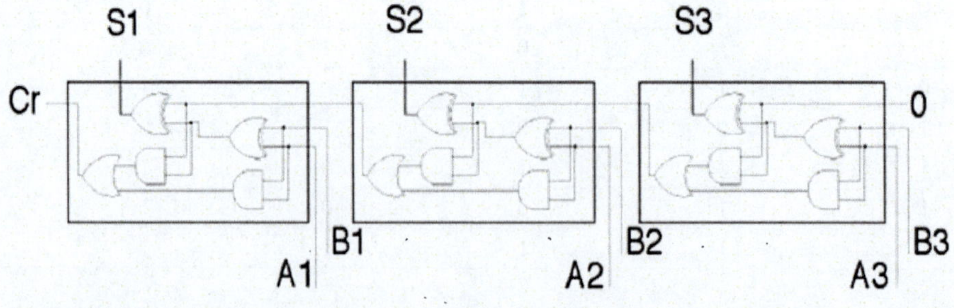

Fig. 2.16 Graphical representation of the three-bit full adder. Result may take up to four bits including the Cr bit

2.6 Logic Gates Made by Logic Switches

In the previous paragraph, we have seen how to do computation using logic gates. In this paragraph, we will discuss how to use logic switches to make logic gates. However, before focusing on logic switches again, it is useful to note that we do not need to find a switch combination for making every logic gate. In fact, it can be demonstrated that there exist families of logic gates, called *universal logic gates*, that can be used as bricks in a building. By combining together these *universal logic gates,* we can build any logic gate and, thus, perform each and any computation that we want.

The NAND gate is one of such *universal logic gates.* It obeys the following simple rule:

0 NAND 0 = 1

0 NAND 1 = 1

1 NAND 0 = 1

1 NAND 1 = 0

In brief, B1 NAND B2 = NOT (B1 AND B2). It is usually represented by the symbol in Fig. 2.17.

By assembling together NAND gates, we can make all the previously seen logic gates. As an example (Fig. 2.18):

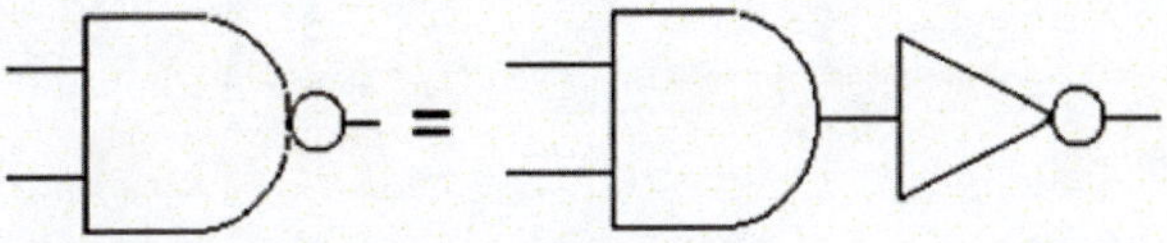

Fig. 2.17 Graphical representation of the NAND logic gate

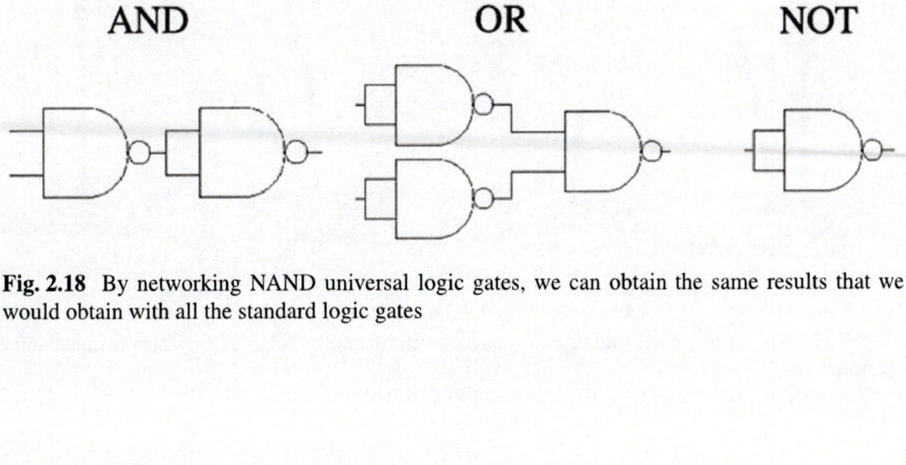

Fig. 2.18 By networking NAND universal logic gates, we can obtain the same results that we would obtain with all the standard logic gates

Clearly, the advantage here is that we can avoid considering all the different kinds of logic gates: we only need to focus on making the NAND gates. To this aim, we need to find a combination of logic switches that perform exactly like a NAND gate.

To date, we have considered only logic switches with one input, in the form of a force and one output in the form of the state of the logic switch. We could simply generalize this approach by allowing more than one force and considering the effect of the resultant of the forces, or in view of a possible combination of logic switches, we could take a different attitude. It would be useful to connect the output of a switch to the input of another switch. With the present configuration, this is not possible because the input is a force and the output is a physical state. Let us try to take a different attitude and consider the arrangement of three switches, as in Fig. 2.19. To fix our ideas, we will consider combinational logic switches in the form of mechanical cantilevers connected by a rope. The position of the two leftmost cantilevers represents the input states, while the position of the rightmost cantilever represents the output state. If we assume that the vertical position represents the physical state S_0 (logic state "0") and the bended position represents the physical state S_1 (logic state "1"), then we have that our arrangement realizes the logic functioning of the NAND gate, as it is easy to verify in Fig. 2.19.

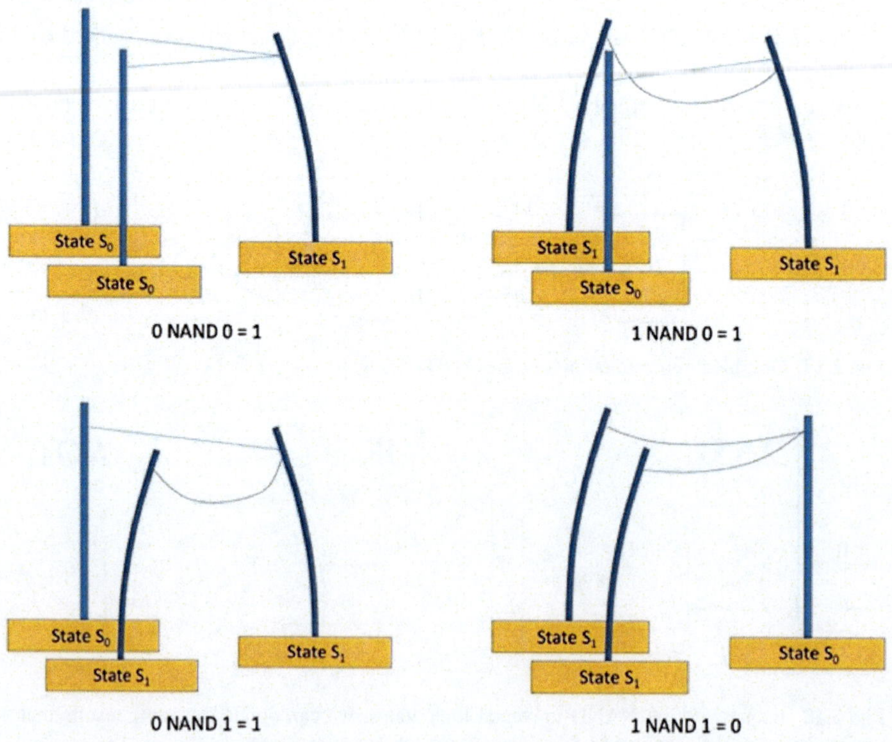

Fig. 2.19 NAND universal logic gates realized with binary switches in the form of mechanical cantilevers

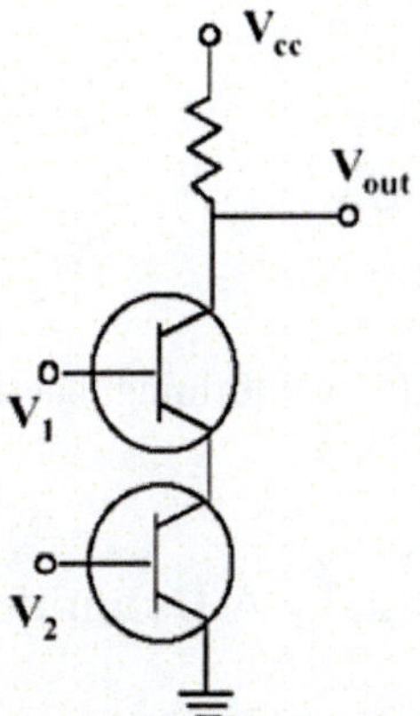

Fig. 2.20 A combination of transistors can be used to make a NAND gate

It should not be overlooked that, even if the three cantilevers that embody our logic switches look the same, their role in the logic gate is not the same. Specifically, the input switches play the role of the force that changes the physical state of the output cantilever and thus the output state. Nevertheless, it is clear that each cantilever represents, by itself, the physical realization of a combinational logic switch, as discussed earlier.

Another, more practically interesting realization of a NAND gate can be obtained by using combinational logic switches in the form of transistors. In Fig. 2.20, we present a combination of two cascaded transistors. This configuration, as mentioned before, has the advantage that the input states and the output states have the same physical nature, i.e. voltages. Here, the two input states are represented by the voltage applied at V_1 and V_2 that corresponds to the *base* terminals of the two transistors. The output state is represented by the voltage at the *collector* of the upper transistor, V_{out}, while its *emitter* is connected to the *collector* of the lower transistor.

If we apply the same convention as before, state S_0, i.e. voltage below a given $V_{th} > 0$ corresponds to logic state "0", state S_1, i.e. voltage equal to V_{cc} corresponds to logic state "1", we can easily verify that the following rules apply:

$$V1 < V_{th} ; V2 < V_{th} \rightarrow V_{out} = V_{cc}$$

$$V1 < V_{th} ; V2 > V_{th} \rightarrow V_{out} = V_{cc}$$

$$V1 > V_{th} ; V2 < V_{th} \rightarrow V_{out} = V_{cc}$$

$$V1 > V_{th} ; V2 > V_{th} \rightarrow V_{out} = 0$$

Or

$$0 \text{ NAND } 0 = 1$$

$$0 \text{ NAND } 1 = 1$$

$$1 \text{ NAND } 0 = 1$$

$$1 \text{ NAND } 1 = 0$$

Confirming that this configuration of transistors realizes a NAND gate.

2.7 A Dynamical Model

As we have seen, logic switches are indeed the basic component of digital computing. In fact, they can be employed to represent stored bits (sequential switches) and, at the same time, they can be connected to perform as logic gates (combinational switches) and do computation.

If we want to address the interesting questions raised in paragraph 1.5 about the physical properties of computing, it seems clear that we need to start with developing a physical model of the logic switches [1]. This can be done by assuming that the switch is a physical system that obeys the laws of physics. By the moment that we are mainly interested in the change over time of state of the switch, it would be wise to look for a dynamical model, i.e. a mathematical representation of the causes and of the effects of the state change. This can be done in different ways.

A simple one is to assume that the state can be represented by a continuous variable $x(t)$ that accounts for a one degree-of-freedom system, subjected to a confining potential $V(x)$ and to an external force F_{ext}. The two states will be identified with respect to the values that such a variable assumes over time: e.g. S_0 when $x(t) < x_{TH}$ and S_1 otherwise.

If we let us inspire by the cantilever realization, we could use the *harmonic oscillator* model to represent the functioning of a combinational switch. In this case, the dynamics is described by a second-order differential equation, as initially suggested by Isaac Newton:

$$m\ddot{x} = -\frac{dV}{dx} + F_{ext}$$

with $V(x) = a\frac{1}{2}x^2$ that plays the role of potential energy. And thus, we have an equation of motion:

$$m\ddot{x} = -ax + F_{ext}$$

Here, the switch dynamics is assimilated to the motion of a point particle subjected to a quadratic potential (see Fig. 2.21) and an external force. The role of the force is

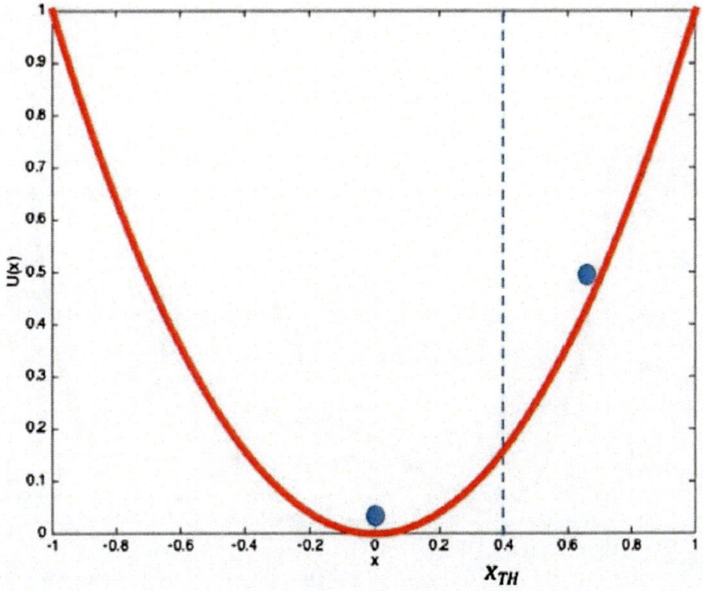

Fig. 2.21 Energy landscape for the point particle dynamics associated with a combinational switch. The presence of the threshold x_{TH} allows for a clear definition of the two states S_0 ($x < x_{TH}$), i.e. point particle at the left of the threshold and S_1 ($x > x_{TH}$), particle at the right of the threshold

to change the position of the point particle above or below a given threshold x_{TH} that marks the separation between the two states. When the force is null, the particle sits at the minimum of the potential ($x = 0$). If we want to switch state, we apply force F_{ext} and the particle is displaced at a position $x > x_{TH}$.

A similar description can be adopted also for the sequential switch. However, in this case we need to consider the specific character of the switch. At difference with the combinational switch that has a single stable state S_0, being S_1 not stable once removed the force, the sequential switch has two stable states: the switch can stay in both S_0 or S_1, once the force is removed. This condition is accomplished by using a different potential $V(x)$. Instead of the well-known quadratic potential of the harmonic oscillator, we need to introduce here a bistable potential that allows for two equilibrium positions.

By keeping the same dynamical equation, we consider now a potential energy in the form of

$$V(x) = -a\frac{1}{2}x^2 + b\frac{1}{4}x^4$$

That gives an equation of motion:

$$m\ddot{x} = \mathrm{ax} - \mathrm{bx}^3 + F_{ext}$$

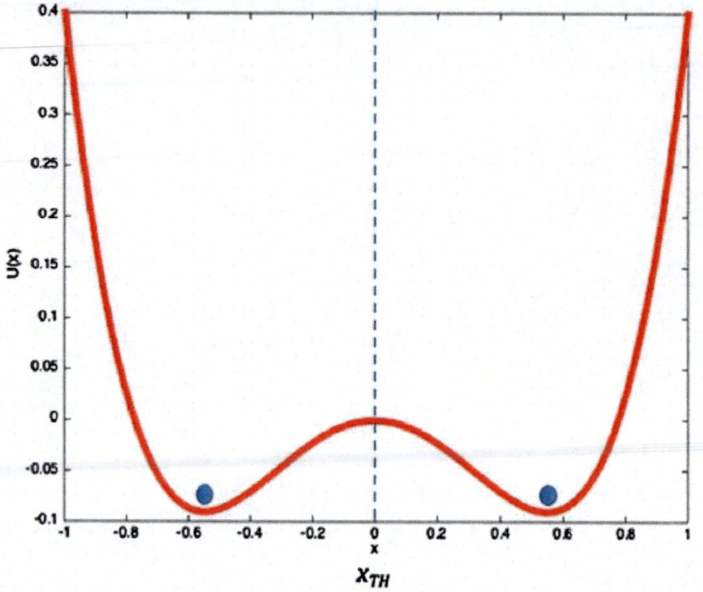

Fig. 2.22 Energy landscape for the point particle dynamics associated with a sequential switch. The two stable states S_0 $(x < x_{TH})$ and S_1 $(x > x_{TH})$ are here easily identified assuming $x_{TH} = 0$

The bistable (quartic) potential is illustrated in Fig. 2.22, where the two states S_0 and S_1 are naturally identified by assuming $x_{TH} = 0$.

In this case as well, the change of state is produced by the application of an external force. However, at difference with the combinational switch, here we need to apply two distinct forces in order to change state: $F_{ext} = F_{01}$ to go from S_0 to S_1 and $F_{ext} = F_{10}$ to go from S_1 to S_0.

Finally, we want to emphasize that the physical model that we are considering is general enough to account for the functioning of a generic logic switch, by the moment that it does obey at the two fundamental requirements that we set when we introduced the binary switches:

- a *physical system* capable of assuming two different *physical states*
- a way to induce state changes in this physical system.

2.8 The Switch

According to our work plan, delineated in paragraph 1.5, in dealing with logic switches we are interested in studying if there are any fundamental physical limits in their functioning. By the moment that all the digital computing activity is performed through them, it is clear that any fundamental limit here will reflect in limitations in computing activities. On the other hand, if we will be able to show that such fundamental limits do not exist, then we expect that this conclusion will also hold for the general computing activity.

To start, we would like to focus on **how much energy** does it take to operate a logic switch. Specifically, we are interested to inquire what is the minimum energy necessary to produce a change of physical state, a situation that goes under the name of *switch event*. In order to produce a switch event, we need to apply an external force. How much work does this force do?

Let us focus on the combinational switch. In order to perform the switch, we apply a constant force $F_{ext} = F$ that simply pushes the point particle above the threshold x_{TH} and up the potential curve, as in Fig. 2.21. If the change in x is Δx, then the external force has performed a work.

$$L = F\Delta x$$

with Δx. Thus, the minimal work required to produce a switch event is $L = Fx_{TH}$. Such a work changes the potential energy of our point particle by the same amount. This potential energy is not lost and can be recovered, once the force is removed. Thus, overall the switch event does require some energy to go from S_0 to S_1, but this energy can be recovered, once the logic switch goes back to its state S_0. This is strictly true only if the sole force acting is an external constant force and no dissipative force, like any friction, is involved. In the presence of friction, clearly our estimate of the work involved is not valid anymore because the energy involved in producing the switch is increased by the amount of energy dissipated by friction:

$$L = F\Delta x + E_d$$

The bad news is that any energy E_d cannot be recovered when the switch goes back to its initial state.

A similar analysis can be conducted for the sequential switch. Due to the symmetry of the potential in Fig. 2.22, the two stable states here are characterized by the same potential energy. To inquire for the amount of energy required for the switch event, we can follow different protocols by applying different forces that change with time.

A first example is represented by the application of a constant force $F_{ext} = F$, as in the previous case of the combinational switch. In general, if a force is conservative, in order to visualize its effect, we might find convenient to include the force in the potential itself and represent the new, effective potential, as:

$$V_e = V - F_{ext}x$$

By the point of view of the dynamics, nothing changes and we go from:

$$m\ddot{x} = -\frac{dV}{dx} + F_{ext}$$

to

$$m\ddot{x} = -\frac{dV_e}{dx}$$

Incorporating the external force into the potential has the effect of showing the role of a changing force as a changing potential. A picture can help to develop intuitive understanding of the switch dynamics.

An important aspect to keep in mind is that, in this representation, the total force acting on the point particle is represented by the negative derivative of the new potential, evaluated in the very position where the particle is. Thus, a particle that sits in a minimum of the effective potential has no force acting on her, while a particle that is close to the maximum slope is subjected at a maximum force. Intuitively, we can look at this representation as the picture of a rolling ball that moves in a gravitational potential.

In Fig. 2.23, we show how the potential is modified by the addition of a constant force $F_{ext} = F$. Here, the role of a positive F is to tilt the bistable potential towards right. If F is small enough, the potential still retains its bistable nature (panel b in Fig. 2.23), and if the force is applied slowly enough, the point particle still sits near the minimum. If the force is not applied slowly enough, some impulse is transferred to the particle that acquires some kinetic energy. This makes the particle oscillate around the minimum. On the other hand, if the applied force is large enough, the bistable potential is strongly tilted and we are left with only one of the two stable states (panel c in Fig. 2.23). The particle is now pushed towards right and will end oscillating around the only minimum that is now present in the potential. In the presence of any friction, the oscillation will eventually dye out and the particle will stay at the minimum. At this point, we can remove the force and let the potential recover its bistable character (panel d in Fig. 2.23). The point particle is now at rest at the right minimum of the potential, and the switch from S_0 to S_1 is completed.

The minimum work required to perform this switch is easily computed by considering that in order to overcome the potential barrier we need to provide the system with suitable potential energy that, in principle, could be recovered once it is transformed in kinetic energy at the bottom of the rightmost well. In the presence of friction, on the other hand, all or part of this energy is lost, in a way similar to what we have seen for the combinational switch. One of the most common kinds of friction is the viscous damping whose intensity is proportional to the velocity of the particle. In this protocol, due to the elevated slope of the potential, we should expect that the velocity of the particle during the switch event is non-negligible and thus it will be the amount of dissipated energy E_d.

Can we imagine a different kind of protocol for the switch event that allows for a lower velocity and thus a lower friction? Yes.

As an example, let us look at Fig. 2.24 where, instead of a applying a simple constant force, we apply a position-dependent force, whose effect is to lower the potential barrier:

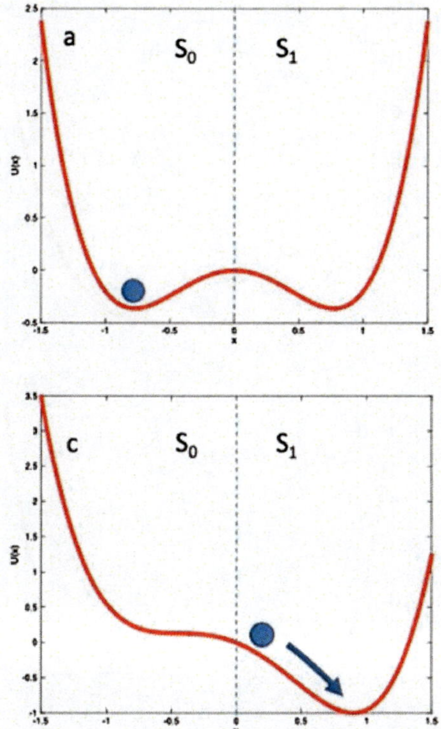

Fig. 2.23 Switch protocol (for the switch S_0 to S_1) based on the application of a constant force F for a certain time. At time t = 0 (panel a), the point particle sits at the left minimum when a constant force F is applied. The effect of the force is to tilt the potential (panel b). If F is strong enough to overcome the barrier height ΔV, the particle will move towards the right (panel c). After some time, once that the kinetic energy acquired is recovered or dissipated, the particle will sit at the right minimum. At this point, the force F can be removed and the switch is completed (panel d)

$$F_{ext} = -\frac{1}{6}ax^3$$

The effective potential becomes:

$$V_e(x) = b\frac{1}{4}x^4$$

In Fig. 2.24 (panel b), we present the situation with the changed potential. Now, the point particle has moved towards the centre of the flat symmetric potential. At this point, in order to switch to the S_1 state we need just a little tilt of the potential towards right. This can be done by applying an additional constant force F_0

$$F_{ext} = -\frac{1}{6}ax^3 + F_0x$$

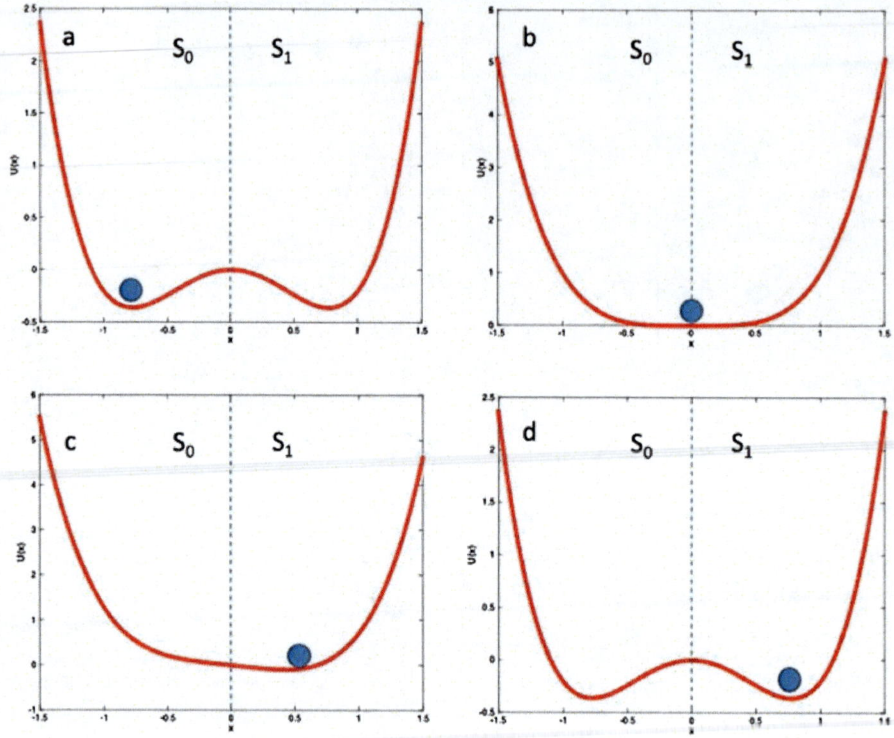

Fig. 2.24 Switch protocol (for the switch S_0 to S_1) based on the application of a position-dependent force and a constant force, for a certain time. At time $t = 0$ (panel a), the point particle sits at the left minimum when a proper position-dependent force is applied to lower the potential barrier (panel b). Now, being the barrier height $\Delta V = 0$, even a small force F_0 is capable of pushing the particle towards the right (panel c). After some time, once that the kinetic energy acquired is recovered or dissipated, the particle will sit at the right minimum. At this point, the force F_{ext} can be removed and the switch is completed (panel d)

and let the particle reach the local minimum (panel c). Now, we can finally remove all the external forces that we added and restore the original potential (panel d).

At difference with the previous protocol, in this case, we can use a constant force $F_0 \ll F$; thus, we will have a velocity of the particle much smaller than before and, accordingly, will be much smaller also the dissipated energy.

What seems to emerge from the discussion of the switch event, both for the combinational and for the sequential switch models, is that the minimum energy required to produce the switch event is represented by the energy dissipated through friction that we have called E_d.

Can we reduce it? Can we make it disappear? In order to answer these questions, we need to understand a little better what *friction* is and where does it come from. This is the topic of the next chapter.

References

1. V. Zhirnov, R. Cavin, L. Gammaitoni, Minimum energy of computing, fundamental considerations, in *ICT-Energy-Concepts Towards Zero-Power Information and Communication Technology*, IntechOpen, 2014

Chapter 3
What is Friction

It is important to realize that in physics today, we have no knowledge what energy is.
Richard Feynman, The Feynman Lectures on Physics, I, 4–1

Abstract In the previous chapter, we laid the foundation for the study of the physics of digital computing systems. In particular, we started inquiring on some fundamental questions about the operating limits of these systems. Given the widespread diffusion of today's computing devices, and the importance that energy consumption plays in our society, it is natural to ask ourselves what are the fundamental limits in energy consumption. How much can we go down in the energy demand for a computing system? In studying the physical basis of these devices, we have come across a phenomenon that seems to be omnipresent and inevitable: friction. But is it really so? In this chapter, we want to deepen this aspect. We will discover very interesting things.

3.1 Energy Conservation

Friction is a common topic in most of our introductory physics courses. It is usually presented as a possible cause when energy is not conserved, thus we have come to think that friction is a mysterious phenomenon that opposes to conservation of energy. We have also invented an expression: we say that energy *gets dissipated through friction*, as we have done in the previous chapter, in order to justify energy disappearance.

However, as we will see in this chapter, friction is not against energy conservation, nor is related to energy disappearance. Indeed, it has more to do with how the energy is *shared among the various degrees of freedom* (dof) of a physical system. In order to better understand this point, let's start with an example.

Suppose we shoot a bullet into a swimming pool. As shown in some recent action movies, the bullet enters the water with a certain speed and then, progressively, it loses speed and, eventually, it will simply fall and stop at the bottom of the pool.

L. Gammaitoni, *The Physics of Computing*,
https://doi.org/10.1007/978-3-030-87108-6_3

When it initially enters the water, has some potential energy V_b and some kinetic energy K_b, so that its total mechanical energy is $E_b = V_b + K_b$. However, when the bullet is at rest at the bottom, its total mechanical energy is $E_b = 0$. Where is all the energy gone? During its descending motion through the water, potential energy is transformed into kinetic energy while the bullet hits a large number of water molecules. During each hit, some kinetic energy is transferred to the molecules that, in turn, transmits such energy to the surrounding molecules and so on. Eventually, all the energy of the bullet is transferred to the molecules that now possess, on average, a larger kinetic energy. Thus, based on this reasoning it appears that energy, that was initially associated only to the motion of the bullet (3 dof if we consider the motion of its centre of mass or 6 dof if we add rotations), it is now shared with a large number of molecules (order of $6N_A$ dof, where N_A is the Avogadro number, $N_A = 6.02214086$ 10^{23}).

If we look at this phenomenon by a merely macroscopic point of view, i.e. by focusing only on the motion of the bullet, it appears that the bullet has decelerated as if it was subjected to a certain force that opposes the motion. We call this force, *frictional force*. Such a force is a very special one, indeed. In fact, it appears only when there is motion and disappears when the bullet is at rest. Moreover, such a force cannot be derived from a potential function, as for the ordinary forces, like gravity, elasticity and so on. There exist different types of frictional force, such as *dry friction* (between two solid sliding surfaces), *fluid friction* (between two liquid sliding surfaces), internal friction (inside a solid, when there are deformations), *viscous friction* (between a solid and fluid, like in our bullet example), just to mention the most common.

Studies on frictional forces started very early in human history. The invention of the wheel was clearly a way to get rid of dry friction. The ancient Greeks and Romans (Aristotle, Pliny the Elder and Vitruvius) dealt with the effects of friction and Leonardo da Vinci himself did some studies on sliding surfaces. Later on, there were contributions from Galileo, Newton, Euler and many others, including Coulomb. Studies on viscous friction where performed by Stokes (1851) and Reynolds (1866). George Gabriel Stokes, Irish born physicist, developed a law for the analytic form of the viscous friction force F_d, in the presence of a spherical body of radius r moving with velocity v through a fluid of viscosity μ:

$$F_d = 6\pi \mu r v$$

that is now known as Stokes' law or Stokes' drag.

What is interesting is that in any phenomenon where there is friction, there is also the appearance of some heat associated with it. Based on our previous discussion of the bullet motion and on our present understanding of the microscopic nature of matter, we can easily interpret the heat as a manifestation of the increased kinetic energy of the microscopic components of the materials in play. According to the kinetic theory of gasses, the temperature T of a rarefied fluid is proportional to the average kinetic energy of its elementary constituents (equipartition of energy):

$$\frac{1}{2}m\langle v^2\rangle = \frac{3}{2}k_B T \tag{3.1}$$

with k_B, the Boltzmann constant ($k_B = 6.023\ 10^{-23}$ J K^{-1}) and $\langle v^2\rangle$ the average of the squared velocity. Thus, the appearance of the heat is a manifestation of the transfer of energy to the surrounding environment: the more energy is *lost* by the macroscopic system, the more heat *appears*. This phenomenon is so common that it has been used to introduce a *mechanical equivalent of heat*. We will come back to this point later on, in this chapter.

In summary, we can describe friction as a process that takes place in a physical system, through which energy is transferred by a small number of dof to a large number of dof.

What has this to do with binary switches and our minimum computational energy? Before answering these questions, we need to dig a little more into friction and what happens at microscopic level.

3.2 Brownian Motion

As we have seen, a fired bullet that travels through the water of a swimming pool, interacts with the water molecules and gives away part of its energy that becomes eventually kinetic energy of a large number of molecules, and this reflects into the increase of the local temperature of the water.

It is reasonable to ask what is the effect of the water molecules, in turn, on the traveling bullet. On the other hand, we know that, even before the molecules are hit by the bullet, they do possess some kinetic energy: the water is at room temperature nevertheless. How does the kinetic energy of the molecules affect the bullet?

By the moment that the mass of the bullet (approx. 16.2 g for 0.45 Calibre) is very large compared to the mass of a water molecule (approx. $3\ 10^{-23}$ g), we might think that the energy of the bullet is so much bigger, compared to the kinetic energy of a single molecule, that the effect of the impacts of the molecules on the bullets may be neglected. This may be true soon after the bullets enters the water but, as time passes and the energy of the bullet decreases, it becomes closer and closer to the average energy of the molecules. Indeed, by definition of (thermal) equilibrium, in the long term, when the bullet sits at the bottom of the pool, the centre of mass of the bullet will have, on average, the same kinetic energy of a single molecule, due to the fact that the bullet will be at the same temperature of the water.

In this equilibrium condition, we expect that the exchange of the kinetic energy continues on a fairer base: on average the bullet takes and gives the same amount of energy to each surrounding molecule. This continuous exchange of kinetic energy manifests itself as random oscillations of the centre of mass of the bullet. The first scientist to note this situation was probably Jan Ingenhousz, a Dutch biologist, credited for the discovery of photosynthesis that, in 1785, described the irregular motion of coal dust particles on liquid alcohol.

A more robust account of such random oscillations has been proposed by the Scottish botanist Robert Brown. Brown, in 1828, published a work entitled "*A brief account of microscopical observations made in the months of June, July and August, 1827, on the particles contained in the pollen of plants; and on the general existence of active molecules in organic and inorganic bodies*" (Fig. 3.1). The experiments performed by Brown were aimed at showing that the random oscillations observed in pollen grains or other very light particles in a fluid were not due to some special form of microscopic life but were some manifestations of the nature of the fluid. A more precise understanding of these experiments has been advanced by the Italian scientist Giovanni Cantoni, who in 1868 wrote [1]: *brownian motion provides us with one of the most beautiful and direct experimental evidence of the kinetic theory of heat, showing that persistent vibrational state that must persist in solids and liquids even when they are kept at constant temperature.*

The idea advanced by Cantoni is incredibly brave, especially in a time where the existence of atoms and molecules was not generally accepted. It is by no means obvious, in fact, that the microscopic movements that were visible under a microscope of pollen grains or other small pieces of matter of size of about 10^{-4} m, were due to incessant hits by much smaller particle, like water molecules that are as small as 2 10-10-10 m. It is like identify the motion of a balloon as large as a whole mountain, subjected to pushes from ants. The idea was so challenging that it was necessary to wait few more years, until 1905, when A. Einstein and, independently, M. Smoluchowski (in 1906), proposed a convincing physical model for such a phenomenon.

The interesting aspect of Einstein–Smoluchowski approach is that it is aimed at describing the random motion of the grain in terms of a macroscopic observable, the diffusion coefficient, that is expressed as a function of microscopic quantities, like the Avogadro number. A similar result was also published, few months before Einstein, by the Australian scientist William Sutherland, but his work was not recognized until later.

Based on what we have said, the *Brownian motion* can be seen as a bridge between the microscopic world, at scales below 1 μm, invisible with an optical microscope, and the macroscopic one. We can have information on the former, by probing the latter.

3.3 Langevin's Approach

In the following, we briefly review the theory of the Brownian motion by adopting a point of view proposed for the first time by P. Langevin in 1908. This approach, in fact, is more clear and insightful compared to the one adopted by Einstein or Smoluchowski shortly before.

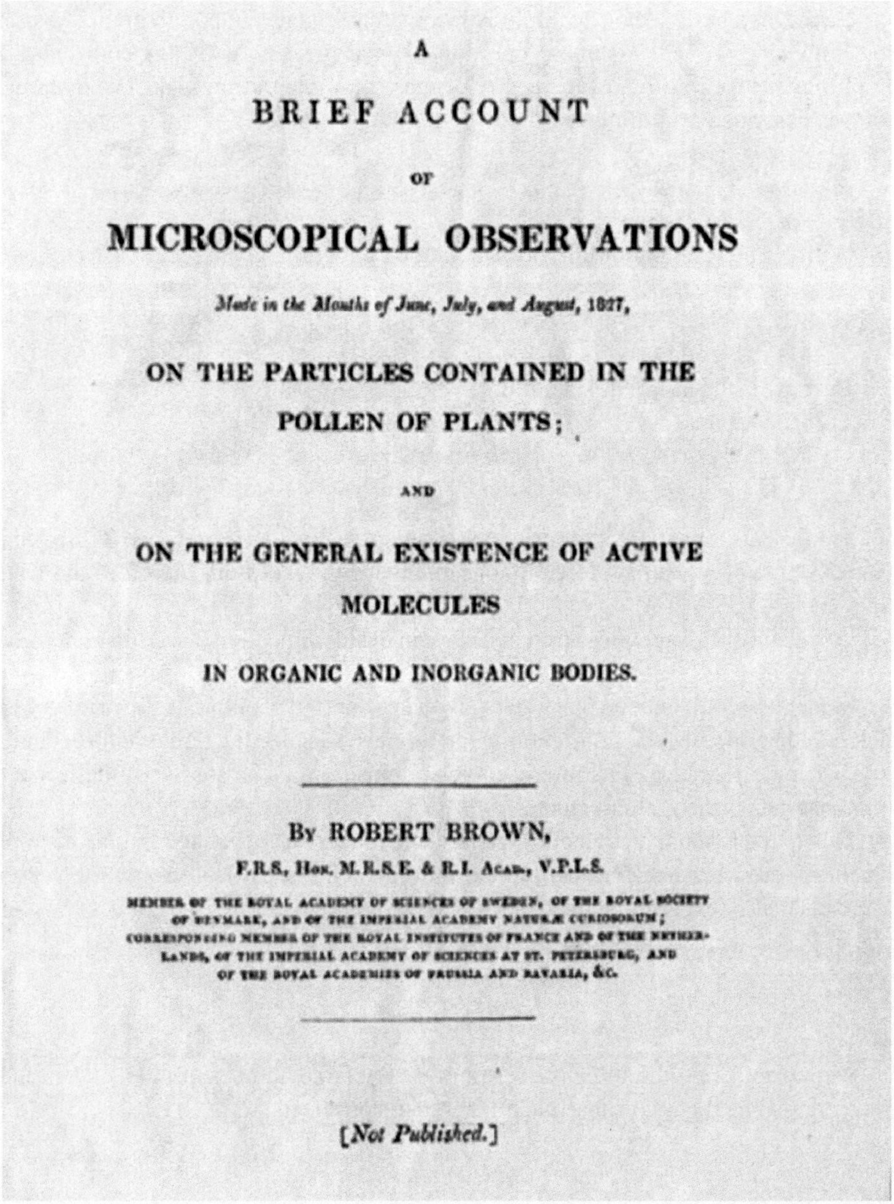

A

BRIEF ACCOUNT

OF

MICROSCOPICAL OBSERVATIONS

Made in the Months of June, July, and August, 1827,

ON THE PARTICLES CONTAINED IN THE POLLEN OF PLANTS;

AND

ON THE GENERAL EXISTENCE OF ACTIVE MOLECULES

IN ORGANIC AND INORGANIC BODIES.

By ROBERT BROWN,

F.R.S., Hon. M.R.S.E. & R.I. Acad., V.P.L.S.

MEMBER OF THE ROYAL ACADEMY OF SCIENCES OF SWEDEN, OF THE ROYAL SOCIETY OF DENMARK, AND OF THE IMPERIAL ACADEMY NATURÆ CURIOSORUM; CORRESPONDING MEMBER OF THE ROYAL INSTITUTES OF FRANCE AND OF THE NETHERLANDS, OF THE IMPERIAL ACADEMY OF SCIENCES AT ST. PETERSBURG, AND OF THE ROYAL ACADEMIES OF PRUSSIA AND BAVARIA, &C.

[*Not Published.*]

Fig. 3.1 Cover page of the work published by Robert Brown in 1828 and entitled: "A brief account of microscopical observations made in the months of June, July and August, 1827, on the particles contained in the pollen of plants; and on the general existence of active molecules in organic and inorganic bodies"

Langevin's basic idea is to apply Newtonian mechanics to pollen grain. For sake of simplicity, we will assume a one-dimensional motion, along the coordinate x, representing the position of the centre of mass of a grain of mass m. The dynamics of x is described according to:

$$m\ddot{x} = F$$

With F, the resultant of the forces acting on the grain, given by

$$F = -F_d + \xi(t)$$

Here F_d is the usual Stokes' drag

$$F_d = 6\pi\mu r v = \gamma\dot{x}$$

that represents the effect of the friction on the grain due to the presence of the liquid water, while $\xi(t)$ is a new force that accounts for the incessant hits from the water molecule.

The beauty of Langevin's approach stays in the identification of two distinct forces that are due to the same physical source: the water. Here the role of the liquid water is, in fact, twofold. On one hand, the water, as a continuous fluid, is the origin of the friction that opposes the grain motion, as we have seen for the bullet before. On the other hand, the water, in its microscopic composition, is the source of the frequent hits that makes the grain fluctuate.

Due to the random character of the hits, the function $\xi(t)$ cannot assume a simple analytic form, but it has to be defined through some statistical properties. By the moment that, as experimentally observed, the average position of the grain is constant, we will assume that $\xi(t)$ is zero-averaged, i.e.

$$\langle\xi(t)\rangle = 0$$

Moreover, being the force the result of a very large number of independent hits, we can assume that two subsequent hits are uncorrelated, i.e.

$$\left\langle\xi(t)\xi(t')\right\rangle = \delta(t - t')$$

The Langevin equation, first example of a stochastic differential equation, can be written as:

$$m\ddot{x} = -\gamma\dot{x} + \xi(t)$$

Multiplying both sides by x:

$$mx\ddot{x} = -\gamma x\dot{x} + x\xi(t)$$

And observing that $\ddot{x}^2 = 2x\ddot{x} + 2\dot{x}^2$, we have:

$$\frac{1}{2}m\ddot{x}^2 - m\dot{x}^2 = -\frac{1}{2}\gamma\dot{x}^2 + x\xi(t)$$

Now, if we take the average over a large number of experimental realizations, we have

$$\frac{1}{2}m\langle\ddot{x}^2\rangle - m\langle\dot{x}^2\rangle = -\frac{1}{2}\gamma\langle\dot{x}^2\rangle + \langle x\xi(t)\rangle$$

In the absence of the stochastic force $\xi(t)$ we would have the simple frictional effect that give us an exponential relaxation towards a steady position, with relaxation time given by $\tau = {}^{m}/_{\gamma}$ that, in standard conditions, is of the order of 10^{-7} s. The hits, on the other hand, have a much higher frequency, with a characteristic time of the order of 10^{-11} s. Thus, based on these considerations (a slow dynamics *vs* a fast dynamics), we can assume that $\langle x\xi(t)\rangle = 0$, being the two phenomena uncorrelated.

If we assume that the system obeys the equipartition of energy (3.1)

$$\frac{1}{2}m\langle\dot{x}^2\rangle = \frac{1}{2}k_B T$$

our equation of motion becomes a standard differential equation for the variance of the position $\langle x^2(t)\rangle$:

$$\frac{1}{2}m\langle\ddot{x}^2(t)\rangle + \frac{1}{2}\gamma\langle\dot{x}^2(t)\rangle - k_B T = 0$$

That, once solved under the condition $\langle x^2(0)\rangle = 0$, gives:

$$\langle x^2(t)\rangle = \frac{2k_B T}{m}\tau^2\left[\frac{t}{\tau} - \left(1 - e^{-\frac{t}{\tau}}\right)\right]$$

When the system has relaxed to an equilibrium state, i.e. for $t \gg \tau$, the transient part goes to zero and we have:

$$\langle x^2(t)\rangle = \frac{2k_B T}{\gamma}t$$

or

$$\langle x^2(t)\rangle = 2D\,t$$

where D is the Einstein–Smoluchowski diffusion coefficient for spherical grains

$$D = \frac{k_B T}{6\pi r \mu}$$

It is important to observe that, in this model, D is a measurable quantity through macroscopic experiments that can provide information on microscopic quantities like the Boltzmann constant.

There is also another, far reaching, observation that can be made. The dynamics of the grain, as we have seen, is characterized by two forces: friction (caused by the viscous nature of the fluid) and noise (caused by the corpuscular nature of the fluid molecules and their random character). These two forces are intimately related because they come from the very same source that is the existence of the fluid itself. However, if we consider the grain+fluid physical system, we can interpret the grain centre of mass just as one of the many dof of our system. The main difference, however, is that this dof is characterized by a very slow dynamics, compared to the other N_A dof. This difference can be overlooked, when the system is at thermal equilibrium, but it is very apparent instead, if we decide to give our grain an initial, strong kick (as in the initial phase of our bullet entering the swimming pool, above). At that point, the grain carries a kinetic energy that is much larger than the average kinetic energy of each dof in the fluid. This excess energy is promptly redistributed through the action of the friction in a process, sometimes called *dissipation*, until the system reaches again the equilibrium state. In this perspective, the dissipation is a process that exists just to let the excess energy to be redistributed through all the dof, in order that any and each of them can keep its random motion (fluctuation) with the same average energy.

As we will see in the next paragraph, such an intimate connection between the dissipative properties and the fluctuations is very profound and, under certain conditions, can be expressed in the form of a mathematical relation: the *fluctuation–dissipation theorem*. However, before getting there we still need one more step.

3.4 Fluctuating Cantilever

All the discussion that we have seen so far is certainly interesting but the connection to our binary switches is still very obscure. Unless we are going to make them of fluid matter, why should we bother about the friction–noise connection?

In order to illustrate this point, let's start with an experiment. Let's suppose that we have built a pendulum, like the one commonly used in undergraduate laboratory experiments. If we displace slightly the pendulum from the vertical position and let it go, we will observe a motion that is approximately periodic with the centre of mass of the ball performing sinusoidal oscillations.

However, if we measure the amplitude of such oscillations, we will note that it decreases with time, according to a law much similar to the exponential law observed for the relaxation of the pollen grain. If we keep observing for a significantly long

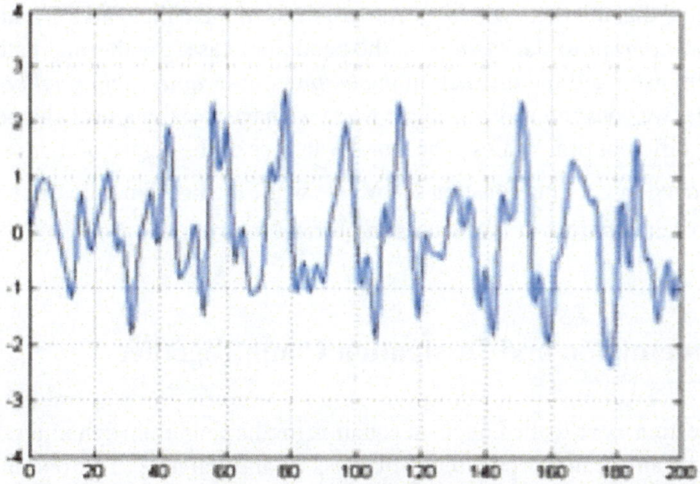

Fig. 3.2 Position of the tip of a mechanical pendulum versus time (arbitrary units)

time (much larger than the relaxation time), the pendulum will appear to have reached a steady position. However, if we observe the position of the centre of mass with a very sensitive measurement system (like an optical interferometer, for example), we note that it is not steady at all, but it performs small oscillations characterized by a significant degree of randomness, much similar to those reported in Fig. 3.2. If we chose a pendulum with mass m = 1 kg, length l = 1 m, we will have oscillations with amplitude such that $\sqrt{\langle x^2 \rangle} = 2\ 10^{-11}$ m.

Where do the observed random oscillation came from? Easy to say: the mass of the pendulum behaves like the bullet in the previous paragraph. It gives away its kinetic energy to the air molecules surrounding the pendulum and finally it reaches equilibrium with the gas.

Now we do something subtle, whose outcome will amaze you: we put our pendulum inside a vacuum chamber and take all the air out. We repeat the experiment and start with the oscillating pendulum. Quite surprisingly, after some time (much longer that before) also in this case the pendulum appears to slow down, until it reaches an apparently steady position. However, a careful observation shows that, also in this case, it is still performing the tiny random oscillations that we observed before in the presence of air. Where do these random oscillations came from now?

The answer is really beautiful: they come from the system itself. The centre of mass of the pendulum behaves as a single dof that interacts, i.e. exchanges kinetic energy, with all the rest of dof (in a number of the order of N_A) that make the real pendulum (wire and mass, altogether).

This is incredibly amazing: in a macroscopic object, i.e. an object made by a large number of dof, we can separate the slow dynamics of one (or few) of its dof from the fast dynamics of the remaining $N_A - 1$ dof. This large reservoir of dof represents what is technically called the *thermal bath*. Such *thermal bath*, in the examples of the bullet in the pool and of the Brownian motion of the grain, was represented by the water liquid that interacts with the centre of mass of the bullet and the grain and was, somehow, *external* to the system. In the pendulum case, the thermal bath is part of the system itself and the distinction *fluctuating observable* vs *thermal bath* is just a model that we conveniently introduce when we have two dynamics characterized by two very different time scales. The slow dynamics of the centre of mass swinging is characterized by the time constant set by $\sqrt{\frac{l}{g}}$, while the fast dynamics is characterized by the correlation time of the fluctuating force $\xi(t)$.

3.5 Fluctuation and Dissipation Come Together

It is interesting to write the Langevin equation for the pendulum dynamics. In Fig. 3.3, we present an idealized pendulum of mass m and length l. The one-dimensional dynamics of the angle $\theta(t)$ can be expressed as:

$$ml\ddot{\theta} = -\gamma l\dot{\theta} - mg\theta + \xi(t)$$

where we have assumed that periodic oscillations have amplitude small enough that $\sin\theta \simeq \theta$. Here the centre of mass position is completely characterized by the angle θ. This is indeed the same equation that we would have considering the well-known harmonic oscillator:

$$m\ddot{x} = -\gamma\dot{x} - m\omega_0^2 x + \xi(t)$$

With $x = l\theta$ and $\omega_0 = \sqrt{g/l}$. The interaction with the thermal bath, as for the Brownian particle, is accounted for by two forces: a frictional term, represented here by a viscous damping proportional to the velocity, with damping constant γ, and a stochastic term $\xi(t)$, that is Gaussian-distributed, zero-averaged and δ-correlated.

Under the linear approximation that we applied, and under the additional hypothesis that the system, at thermal equilibrium, obeys the energy equipartition (3.1), it is possible to establish a mathematical relation that connects the two forces that originate from the presence of the friction. As we have anticipated in Sect. 3.2, this relation goes under the name of *fluctuation–dissipation theorem* and can be expressed as:

$$\langle \xi(t)\xi(t') \rangle = 2k_B T\gamma\delta(t - t')$$

or, what is equivalent:

$$\langle \xi(t)\xi(t - \tau) \rangle = 2k_B T\gamma\delta(\tau)$$

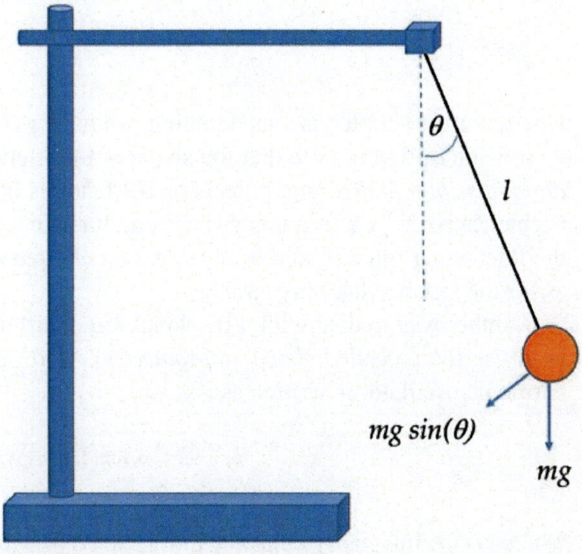

Fig. 3.3 Simple pendulum

This relation has been established as a theorem, within the linear response theory (that satisfies the equipartition of the energy among all the degrees of freedom) initially by Harry Theodor Nyquist (1889–1976) in 1928, and demonstrated by Callen and Welton in 1951 (Fig. 3.3).

Its main content is the quantitative link between the two forces that we have introduced, starting from the Brownian motion, and discussed so far. The existence of such a relation says that we are not free to choose at will the intensity of the fluctuating force and of the damping force, but it does exist an intimate connection between the two: if we change one, also the other one is affected and that's because both arise from the same physical origin.

The autocorrelation function of the fluctuation $C(\tau) = \langle \xi(t)\xi(t-\tau)\rangle$ is a delta function and is independent on t for all the stationary processes, just as we assumed for the Brownian motion. This kind of fluctuation is usually called *white noise* because its frequency spectrum $S(\omega)$, expressed through the Wiener–Khinchin theorem [2], as the Fourier transform of the autocorrelation function, is constant:

$$S(\omega) = \int_{-\infty}^{+\infty} C(\tau)e^{-i\omega\tau}d\tau = \langle \xi^2 \rangle$$

In general, the *fluctuation–dissipation theorem* holds true also for frictional forces that are different from the simple viscous damping F_d. As a matter of fact, most frictional forces have a dependence from dynamical quantities that are more complicated than the simple proportionality to velocity. In general, we can assume that the frictional term γ, instead of being a simple constant, is a function of time $\gamma(t)$, whose effect in the dynamics carries also memory of the previous history of the system. A general way of writing such a frictional term is the following:

$$\int_{-\infty}^{t} \gamma(t-\tau)\dot{x}(\tau)d\tau$$

where, for the simple viscous damping, we have $\gamma(t-\tau) = \gamma\delta(t-\tau)$.

It is interesting to note that the shape of the frictional term, thanks to the *fluctuation–dissipation theorem*, reflects on the features of the fluctuating force, that now is characterized by a frequency spectrum that is not constant anymore. In this case, the fluctuating force is said to represent a *coloured noise*, instead of a white noise, as for the viscous damping case.

Another way to deal with a frictional force, different from the viscous damping, is to use the so-called *Zener approach* [3]. Here, the equation of motion for the harmonic oscillator is written as:

$$m\ddot{x} = -m\omega_0^2(1+i\phi)x$$

where i is the imaginary constant and ϕ, called *loss angle*, is a function of frequency. This equation is not a stochastic differential equation, as before, because now we have an imaginary component in the dynamics. By the moment that $x(t)$ is a real quantity, you can understand that this makes things a little more complicated. Be that as it may, this approach is quite common among those who deal with a kind of frictional force that arise in solid materials. As we have seen for a real pendulum, once we have taken out the air, the suspension wire, that is subjected to bending and stretching, shows a dissipation that comes from the wire itself (Fig. 3.4). Here the bending of the physical wire produces a deformation of the solid that it is usually subjected to friction.

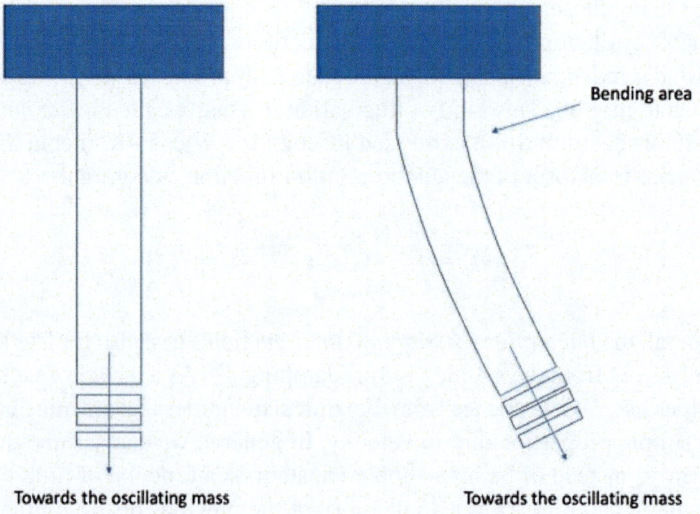

Fig. 3.4 Pendulum wire in vertical position (left) and at the point of maximum oscillation (right)

There are at least two different mechanisms that, in this case, may contribute to dissipation. The first one is called *internal friction* and is associated with the movement of defects, like vacancies and dislocations [4], inside the crystal lattice and is usually modelled with a constant loss angle $\phi(\omega) = \phi_0$.

The second one is called *thermoelastic damping*, and is due to the fact that, when a wire is bended, on one side there is compression while on the on the opposite side there is dilatation (alternatively as the oscillation goes on). This asymmetric condition causes some thermal transport from the compressed side towards the dilated side and this transport is the source of some dissipation, i.e. some redistribution of the kinetic energy through heat diffusion. For the *thermoelastic damping,* it is usually assumed a loss angle with a frequency dependence expressed by:

$$\phi(\omega) \propto \frac{\omega}{1 + (\omega\tau_0)^2}$$

where τ_0 is a thermal transport relaxation time, typical of the wire material.

Finally, by the point of view of the *loss angle*, the viscous damping is, instead, characterized by a $\phi(\omega) \propto \omega$.

If one is interested building high precision instruments, like for example micromechanical resonators for sensing application, it is important to carefully characterize the frictional forces behaviour, because fluctuations generated by those forces might limit the instrument sensitivity. Such a limiting source is sometimes called *thermal noise*.

This is the case, for example, of the gravitational wave detectors, where *thermal noise* of the structures associated with the optics (mirrors and their suspension wires) of the laser interferometer provides an important contribution to limit the detector sensitivity [5].

3.6 Thermodynamics for Everybody

According to Sect. 3.1, the best way to understand friction is to describe it as a process that takes place in a physical system, through which energy is transferred by a small number of dof to a large number of dof. When this happens, the temperature of the system increases and we say that some heat has been produced at the expenses of some work done on the system. In fact, for a conservative system, the change in kinetic energy ΔK is equal to the amount of work L done on/from the system. Thus, it is straightforward to conclude that, when some kinetic energy seems to disappear, some heat Q appears such that:

$$\Delta K = L = Q$$

This possibility to change some work completely into heat, was initially formulated by Benjamin Thompson, an American-born scientists that conducted, mainly in UK and Germany, a quite adventurous life. Thompson noticed that the work done by boring cannons at the arsenal in Munich was always equivalent to a given quantity of heat produced. He published his observations in 1798 in a work entitled "*An Experimental Enquiry Concerning the Source of the Heat which is Excited by Friction*". Since then, the idea that heat was a form that energy can assume, spread quite generally, providing us with one of the first generally recognized expressions of the conservation of energy. Thanks to the work of scientists like Julius Robert von Mayer (1814–1878) and subsequently, James Prescot Joule (1818–1889) and Hermann Ludwig Ferdinand von Helmholtz (1821–1894), the conservation of energy, for systems where the potential energy does not change and heat is involved, can be expressed as follows:

$$\Delta U = Q - L$$

This expression is also known as the **first principle of thermodynamics**.

By the moment that putting in close contact two bodies their temperatures eventually become equal, it derived the idea that heat could be exchanged between the two. In this sense, heat and work appear like two distinct ways to change the total energy of a system when this interacts with the outside (see Fig. 3.5). For historical reasons, the work is considered as a positive quantity when it is done by the system (outgoing) and heat is considered positive when entering the system (incoming).

Based on what we have seen so far, we understand that is always possible to transform work into heat. Friction is a good example of this transformation. How about the reversal? Is it possible to transform heat into work?

The idea is the following: we provide some heat to the physical system and in exchange we get the same amount of energy in the form of work. Indeed, this is a great idea because we have plenty of energy available in the form of heat, think for example at the air, or the sea. We could put our system in contact with such large reservoirs of heat and get work to make our machines go. Can we do it?

The answer is yes, but there are some limitations. Such limitations are very important from a practical point of view because they set some constraints to the functioning

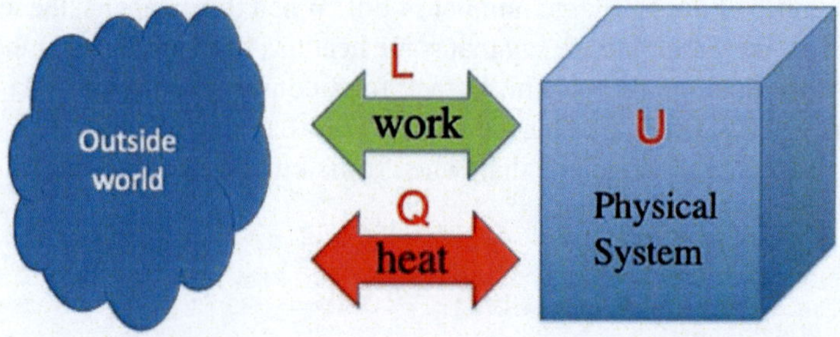

Fig. 3.5 A physical system exchanges energy with the outside, in the form of work L and heat Q

of the so-called *thermal machines*, like coal-powered machines that played a major role during the industrial revolution. The content of these limitations constitutes the **second principle of thermodynamics**.

According to William Thompson (1824–1907), Lord Kelvin, probably the most eminent scientist in this field, the situation is the following:

> No process is possible in which the sole result is the absorption of heat from a reservoir and its complete conversion into work.

A different but equivalent formulation is due to Rudolf Clausius (1822–1888):

> No process is possible whose sole result is the transfer of heat from a body of lower temperature to a body of higher temperature.

In both cases, we see that there are limitations to how much work can we get from a given amount of heat. This principle is based on many experiments performed during the eighteenth century and nineteenth century: no matter how well we built our thermal machines, we were not able to get, as work, all the energy available as heat.

A question rises spontaneously: what happens to the amount of heat that we cannot change into work? The answer to this question has been provided by Clausius in what is presently called the *Clausius theorem*. In order to understand this point, let's go back to the Kelvin's formulation of the second principle and ask again: if it is not possible to absorb *heat from a sole reservoir and* completely transform into work, what do we do with the remaining heat? The answer is: we have to give the heat to another reservoir that is at lower temperature (due to the Clausius formulation). In fact, this is exactly what has been proved by the French scientist Sadi Carnot in 1824, when he was trying to idealize the functioning of a thermal machine. He discovered that it was necessary to operate it between two different temperatures. Now, as we discovered, when we give the residual heat to a lower temperature body, this will increase its temperature and thus the thermal agitation of its elementary components. This is a phenomenon similar to the dissipation we studied before: even more dof will become interested in sharing a given amount of energy.

Clausius was smart enough to figure out a way to quantify this phenomenon and introduced a state function represented by the ratio between the amount of heat Q exchanged with other sources and the value of the temperature T of these sources. He showed that if we perform a cyclic transformation, we have:

$$\oint \frac{dQ}{T} = 0$$

Provided that during the process, the heat exchanges are performed by maintaining constant equilibrium inside the system. In practice by performing the exchanges very slowly and between systems that are at very close temperatures (a process that nowadays we call *reversible transformation*). If this condition is met, then Clausius equation holds and it mathematically implicates that it exists a physical quantity S, that he called *entropy*, such that:

$$\int_A^B \frac{dQ}{T} = S(B) - S(A)$$

where A and B are two states of the system.

If the transformation is not reversible (i.e. is *irreversible*), instead, the amount of heat exchanged satisfies this relation:

$$\int_{A irr}^{B} \frac{dQ}{T} \leq S(B) - S(A)$$

If we consider an infinitesimal transformation, we have:

$$\frac{dQ}{T} \leq dS$$

or

$$TdS \geq dQ$$

where the equal sign holds during a reversible transformation only.

Let's suppose that we have a transformation where no heat nor work is exchanged and it is performed out of equilibrium: it is called an adiabatic spontaneous transformation (like the free expansion of a gas). In this case, we have $Q = 0$ and:

$$dS > 0$$

Thus, we conclude that during a spontaneous adiabatic transformation, the entropy always increases. This is sometimes considered another formulation of the **second principle of the thermodynamics.**

The quantity $T\ dS$ can be interpreted as the amount of energy that cannot be transformed into work but has to be given to the lower temperature reservoir to increase its entropy by dS. Thus, to answer our previous question, if we have a given amount of thermal energy U, we can use part of it, that we will call *Free Energy* F, according to the definition introduced by the German scientist Hermann von Helmholtz (1821–1894), to do work, while the remaining part in the measure of quantity $T\ dS$ will be taken by the change in system entropy, according to:

$$dF = dU - TdS$$

In general, the equilibrium condition for our thermodynamic system is represented by arrival point of the spontaneous evolution, characterized by the tendency to reach a minimum of the free energy. This is thus characterized by the tendency of the energy to reach a minimum and the entropy to reach a maximum.

In summary, what we have learned so far is that we can transform work into heat, just like when we have friction but, even if the total energy is conserved (first principle), to go back to work is not always possible because the possibly required change of entropy may stand in our way (second principle). We should remember that it is always possible to increase entropy at a net zero energy expense (like in a free expansion of a gas). Vice versa, decreasing entropy it does require a net energy expense equal at least (reversible transformation) to $|T\Delta S|$ energy.

3.7 Energy and Entropy at Small Scales

Even if more than 150 years has passed since Clausius formulated his theorem, the notion of entropy has been for long time and, somehow still remain, a quite obscure concept. A very significant contribution to its clarification has been provided by Ludwig Boltzmann (1844–1906), a great scientist, that cast light on the role of energy and entropy at microscopic scales.

In order to better grasp this meaning, let's start with a simple model [6]: a particle gas in a box with a movable piston.

In Fig. 3.6, we present a graphic representation of such an ideal particle gas. It is made by N hard spheres, all with the same mass m, that can elastically bump into each other and with the walls of the box, including the movable piston of mass $M = Nm$, attached to a spring of elastic constant k. This is the kind of model that is often used in the concept of ideal gas and was very useful, in the nineteenth century, to develop the kinetic theory of gasses. We suppose that N is a very large number, of the order of N_A and, for simplicity, that all the particles have the same velocity v. We can define a temperature T for this ideal gas, based on Eq. (3.1): the larger is the velocity of the particles, the higher is the kinetic energy of the gas and thus its temperature.

Let's consider the interactions between the particles and the piston in two very different situations.

In the first case, we will consider a gas where the particles move coherently towards right, all with the very same velocity (see Fig. 3.7). This is a very unlikely

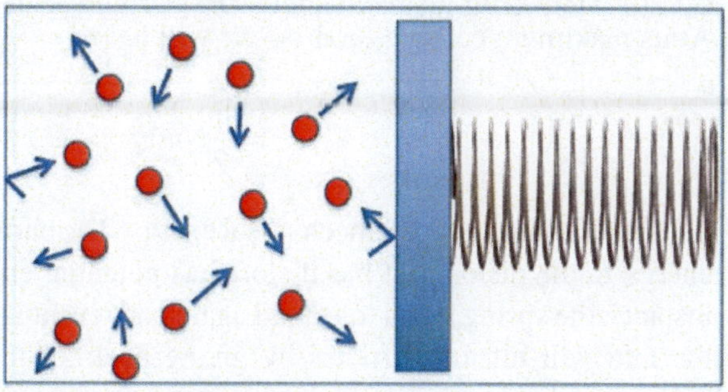

Fig. 3.6 An ideal gas is made by hard spheres (red dots) that can interact through elastic hits with the walls of the box, the piston and each other

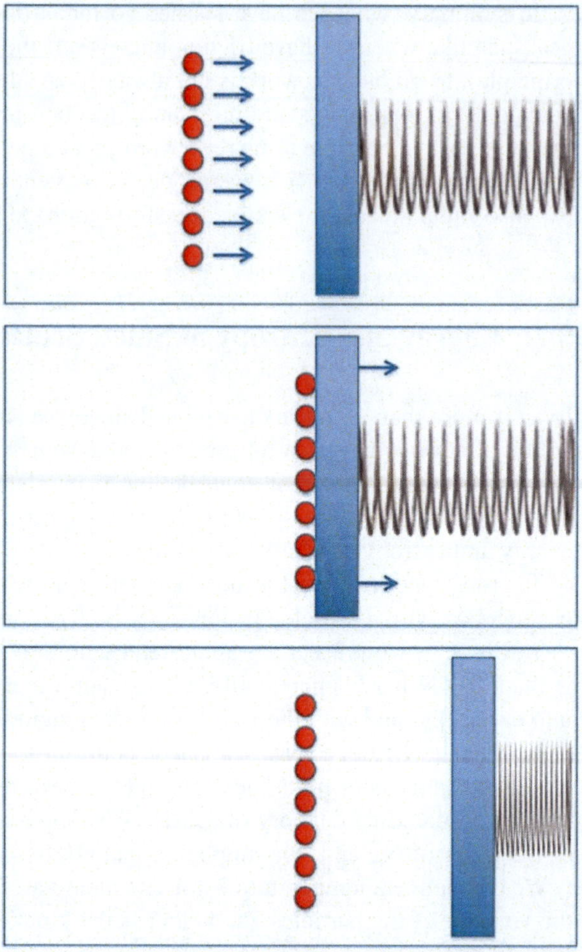

Fig. 3.7 First case: all the particles move coherently with the very same velocity (top). After the collision, the particles stop and the piston starts to move towards right (middle). At the maximum compression, the piston stops and reverse its motion (bottom)

case, indeed. However, it is not forbidden by the laws of physics. What would happen in this case? At a given instant, all the particles collide with the piston and, due to the conservation of the impulse and of the kinetic energy, they simply exchange their velocities. Just after the collision, the particles stop and the piston starts to move towards right with the same velocity v, compressing the spring to a certain extent. At its maximum compression x_M we will have:

$$U = \frac{1}{2}\text{N}mv^2 = \frac{1}{2}kx_M^2$$

With U being the total mechanical energy. The particles have given all their kinetic energy to the piston that has it stored as potential energy of the spring. After some instants, the spring starts to extend again and, eventually, the piston, moving towards the left, will hit the particles by giving back all the energy. Here, we observe a continuous exchange between the kinetic energy of the particles and the potential

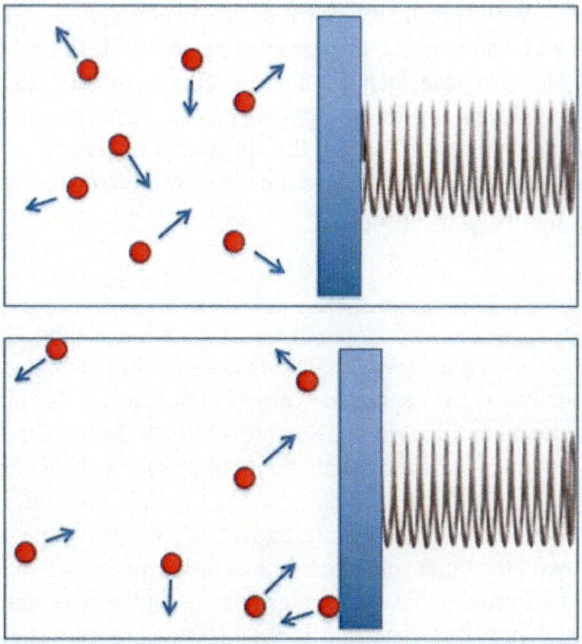

Fig. 3.8 Second case: all the particles are moving randomly around, with the same modulus of the velocity but with different directions (top). The piston gets hit at random times. After each hit, it will show an average spring compression such that its mean potential energy is U/N

energy of the piston that is built by the work done by the elastic force of the spring: there is complete conversion of heat into work (and vice versa).

In the second case, as illustrated in Fig. 3.8, we have a quite different situation. Here all the particles are moving randomly around, with the same modulus of the velocity but with different directions. Based on our experience, this is clearly a most probable configuration for a particle gas. Please note that the energy of the gas is still the same (so is its temperature T), but in this case the piston gets hit at random times and it is thus subjected at random motion with an average compression of the spring such that its mean potential energy is U/N. This is also the maximum work that we can recover from the potential energy of the movable piston.

It is interesting to note that, although the total mechanical energy U is the same in the two cases, in the second case we have no hope of using the greatest part of this energy to perform useful work. Based on our previous discussion, when we introduced the notion of *Free energy*, the quantity that limits our capability of performing work is the entropy. The systems that have the smaller entropy have the larger capability of performing work. It is clearly the first configuration of our system that bears the smaller entropy.

This example helped us to understand how energy and entropy are connected to the microscopic properties of the physical systems. In the simple case of an ideal gas, the system energy is nothing else than the sum of all the kinetic energies of the single particles. We can use the entropy to add a *quality* feature on the energetic content of a system. Two systems may have the same energy, but the system that has the lower entropy will have the "most useful" energy.

With specific reference to our ideal gas, we can say that the energy is associated with "how much" the particles move. While, based on the arrangement of the particles, we have seen that there is also a "quality" of the motion of the particles that is relevant for the entropy. We can say that the entropy is associated with "the way" the particles move, i.e. the specific arrangement or configuration. This concept of "way of moving" was made clear by Boltzmann who proposed for the entropy the following definition:

$$S = k_B ln W$$

where k_B is the already mentioned Boltzmann constant, W is the number of microstates associated with a given state of the physical system and ln indicates the natural logarithm. Sometimes W is also called the "number of possible configurations" and represents the number of ways we can arrange all the particles in the system without changing its macroscopic properties.[1] In the previous example, we have only one way to arrange the N particles so that they are all parallel, aligned and with the same velocity, while in the second case we have very large number of ways of arranging the N particles to be a randomly oriented set of particles with velocity v. Thus, it is clear that in the second case, the value of the entropy is much larger than in the first case (where it is indeed zero).

A more precise way to count the number of configurations is provided by the concept of *phase space*. If we have a particle gas of N particles, the state of the system is described by each particle's position $\boldsymbol{r}$ and momentum $\boldsymbol{p}$, i.e. by $6N$ variables. The value of these variables identifies a point (called representative point) within a $6N$-dimensional space (called *phase space*). The evolution with time of our gas is represented by the motion of the representative point within the phase space. For a thermodynamics system, we can fix some relevant extensive observables, like the total energy U, the occupied volume V, and ask what are the corresponding values that the representative point can occupy within the phase space, compatible with these fixed values of the observables.

From an experimental point of view, each representative point will be characterized by some uncertainty Δr and Δp that will define a tiny iper-cell in the $6N$ iper-space of the phase space, surrounding each point. The volume of such iper-cell cannot be reduced to zero due to the Heisenberg uncertainty principle that sets a lower limit to this volume.

If we call Γ the iper-volume composed by the all iper-cells occupied by the representative point, then the Boltzmann entropy can be defined as:

$$S = k_B ln \Gamma$$

[1] Here we assume that all the microstate are equiprobable. The extension to the more general case with microstates with different probabilities has been proposed by Josiah Willard Gibbs (1839 – 1903).

In the following chapter, we will discuss how to apply what we have learned here to logic switches and their functioning but before that, we will give another look at what we have said about friction.

3.8 Friction at Small Scales

It is interesting to apply the particle gas model that we have introduced in the previous paragraph also to the study of the friction. As we have seen, by friction we mean a process that transfers energy from few dof to many dof. In order to illustrate this process, let's consider the situation in Fig. 3.9, from top to bottom. Here we move the piston to the right and compress the spring by accumulating potential energy. When we release the piston, the potential energy transforms into kinetic energy of

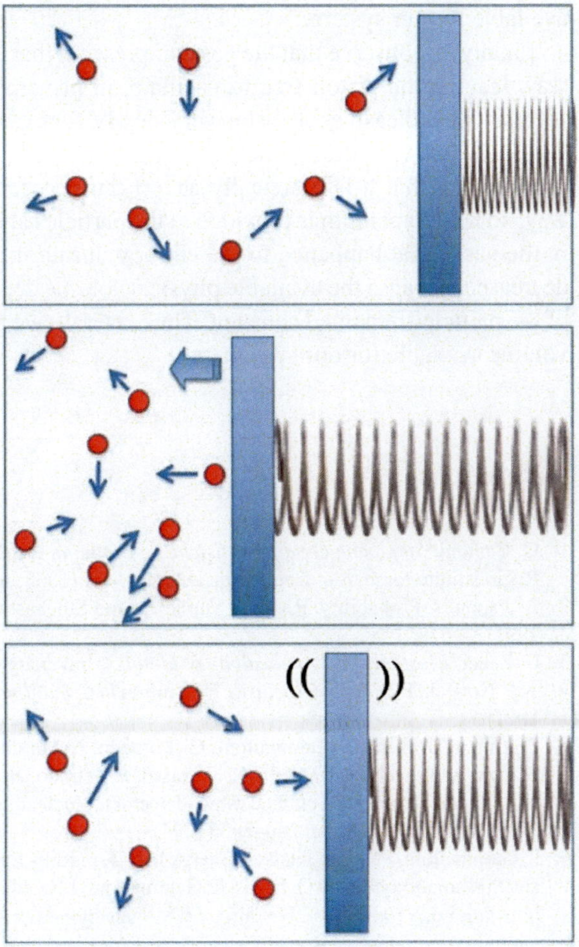

Fig. 3.9 Top: the spring is compressed, and the piston is kept at the right. Middle: the piston is released. The potential energy of the spring transforms into kinetic energy of the piston that moves towards left and hits the particles, transferring to them some of its energy. Bottom: after few oscillations, an equilibrium is reached with the piston oscillating at a constant amplitude x

the piston moves towards left hitting the particles and transferring to them some of its energy. After few oscillations, we observe that the amplitude of the periodic motion of the piston decreases, as a consequence of the friction (viscous damping force) action. The decrease ceases when the oscillation amplitude x reaches an equilibrium value such that

$$\frac{1}{2}mv^2 = \frac{1}{2}kx^2$$

and after that it remains constant. Clearly, this is an irreversible process, where some work has been changed into heat. As a matter of fact, the average kinetic energy of the particles is increased and so the temperature of the gas. The small, residual fluctuations of the piston at equilibrium are due to the fluctuation–dissipation connection that we discussed in 3.5. What happened to the entropy during this process? It clearly increased. In fact, by the moment that the available volume decreased but the average velocity of the particles increased and, now, we have a wider phase space volume available for our system.

Finally, we observe that the gas compression that we have made in an irreversible way, leaving the piston free to oscillate, in principle, could have been made also reversibly by releasing the piston very slowly, for example by pushing against another spring. In this way, during the gas compression, if the piston moves very slowly we can imagine that it is practically at rest during each hit with the particles. In this way, no net momentum is provided to the particles and no kinetic energy is increased in the gas. What happened to the entropy during this reversible process? It clearly decreased, because the available physical volume decreased and the average velocity of the particles remained constant. Thus, overall, we now have a smaller phase space volume available for our system.

References

1. G. Cantoni, *Su alcune condizioni fisiche dell'affinità, e sul moto browniano*, in Rendiconti del Regio istituto lombardo di scienze e lettere, s. 2, I (1868), pp. 56–57
2. A. Papoulis, Probability, Random Variables, and Stochastic Processes, 3rd edn (McGraw-Hill, 1991)
3. C. Zener, *Elasticity and Anelasticity of Metals* (University of Chicago Press, Chicago, 1948)
4. A.S. Nowick, B.S. Berry, *Anelastic Relaxation in Cristalline Solids* (Academic Press, New York, 1972)
5. P. Amico, L. Bosi, L. Gammaitoni, G. Losurdo, F. Marchesoni, M. Mazzoni, M. Punturo, R. Stanga, A. Toncelli, M. Tonelli, F. Travasso, F. Vetrano, H. Vocca, Thermal noise reduction for present and future gravitational wave detectors. Nuclear Instruments and Methods in Physics Research A, Vol. 518, pp. 240–243 (2004)
6. L. Gammaitoni, Fundamentals on Energy in ICT, in ICT - Energy Concepts for Energy Efficiency and Sustainability, ed. by G. Fagas, L. Gammaitoni, J.P. Gallagher, D.J. Paul. IntechOpen, (2017). https://doi.org/10.5772/66973. https://www.intechopen.com/chapters/54033

Chapter 4
The Physics of Small-Scale Switches

Why, these balls bound; there's noise in it.
W. Shakespeare, All's Well That Ends Well

Abstract In the previous chapter, we have learned how to deal with energy dissipation at macro- (thermodynamics) and microscale (dissipation/fluctuation). Here we will apply this knowledge to the physics of binary switches. By the moment that present generation of computing devices are small, we will need all our understanding of the concepts discussed so far to get a realistic and complete description of the physics of these devices.

4.1 Logic Switches as Thermodynamic Systems

In Chap. 2, we have left our discussion of switches with questions on E_d, the energy dissipated through friction during the switch event. Now, after our discussion in Chap. 3, we have a clearer idea of what friction is and how to deal with the energy exchanges through work and heat in a thermodynamic system. But how is this relevant for our discussion of the switch event? After all we are dealing with a dynamical system that can be described by one or few dof as we have discussed in Chap. 3, and they are usually *small* and *isolated* from the outside.

In order to better understand this aspect, we should keep in mind that when dealing with logic switches, even if we focus on a single dof, we have nevertheless to do with physical systems made by a N number of atoms. How large is N? Well, in present-day microtechnology, logic switches can be composed by a number of atoms N that is small compared to N_A but, nevertheless, still in the order of 10^6 at least. This is the case, for example, of the nanodevices like last-generation transistors. However, by the moment that logic switches are usually part of a larger assembly of switches, they cannot be considered truly isolated because they exchange energy with the outside, in the form of work and heat.

L. Gammaitoni, *The Physics of Computing*,
https://doi.org/10.1007/978-3-030-87108-6_4

However, for the sake of discussion, let's suppose that we can consider a single logic switch that is isolated (if necessary including the nearby environment). We plan to address the switch dynamics in a way similar to what we could do with the movable piston and the particle gas, in the previous chapter. There, our system is composed by 3N+1 dof: 3 for each of the N particles and 1 for the movable piston position coordinate x.

We notice that the dynamics of the piston remind us of the dynamics of the pollen grain in the Brownian motion case. $x(t)$ evolves in time much slower that the particles of the gas whose velocity is on average much larger. In this case, we can focus our attention on the single degree x and summarize the role of the remaining 3N dof by introducing the dissipative and the fluctuating force like external forces, as we have done for the Brownian motion and in the pendulum discussion. By the moment that both these forces are necessary to account for the observed dynamics and, by the moment that both are born out of the neglected 3N dof, it comes out that they are connected to each other and the connection is nothing else that the *fluctuation–dissipation relation* that we discussed in the previous chapter. Thus, our equation of motion, instead of being the deterministic *Newton equation*, it will be a stochastic *Langevin equation*, where there is both friction and fluctuations.

This is the approach that we are going to use for the logic switch dynamics [1]; we focus on one dof (the position of the tip of the cantilever, the value of the electric voltage, the value of the magnetization, …) and consider the other 3N dof as the thermal bath that provides friction and fluctuations. This is also the thermal bath that can absorb and provide heat in exchanges with the outside. In this regard, our switch has to be considered as a peculiar thermodynamic system, and changes in entropy and energy have to be properly considered.

4.2 A Digression: Energy Dissipation is an Illusion?

Before dealing with the main problem at hand, i.e. the description of the switch dynamics, we would like to make a brief digression to discuss one aspect of the dissipation of energy in small-scale systems that is somehow puzzling. As we have just seen, in dealing with the piston–gas system, as a paradigmatic example for our logic switch behaviour, we have played a smart trick: we focused our attention on the single dof x and summarized the role of the remaining 3N dof by introducing the dissipative and the fluctuating force like external forces. Thus, the trick we played was to exchange the dynamics of a *not small isolated* system (our switch made by 3N dof) with a *small not isolated system* (one dof of the switch in contact with the thermal bath composed by the 3N-1 dof). Such an approach has different names (adiabatic elimination, coarse graining, …), and it is considered a very useful tools in describing the properties of dynamical systems composed by many dof. Why?

Because the dynamics of a *not small isolated* system can be described in terms of 3N dof by 3N coupled motion equations and, when N is very large (order of N_A) this is a practically impossible approach. Thus, the advantage was to drastically reduce the number of equations of motion (in this case to just 1) but the price we had to pay is the introduction of external forces: dissipation and fluctuation. What we have found is that dissipative and fluctuating effects appear only if we neglect some (usually many) dof through some coarse graining approximation to the system dynamics. In this perspective, the dissipation of energy appears to be only an illusion due to our choice of dynamical description.

On the other hand, in our daily life we are accustomed with the effect of the dissipation of energy to the extent that we use the dissipation to distinguish between forward and backward propagation of time. In fact, if we observe a backward reproduction of a movie of the piston–gas system, we can say immediately that there is something odd: the energy instead of being dissipated from the piston motion to the gas is actually transferred from the gas and accumulated in the piston. In a thermodynamics description, that is just to say that the entropy, instead of being increased during the spontaneous transformation, is miraculously decreased. But if we go back to the whole dynamic description of the 3N dof, there is no dissipation nor odd time evolution because we only see energy exchanges between the many dof involved. Once again it seems that dissipation and entropy increase (and rightful time evolution) are just a consequence of the assumption of a specific dynamic description. In a certain sense, this is actually the case. In this whole dynamic description, there is nothing like the second law of thermodynamics, nothing that forbids the accumulation of energy from a single dof. It is just very unlikely to happen to the very specific dof that we picked. After all it is one out of N dof although one very special to us. On the other hand, if you wait long enough, until the equilibrium is reached, and you watch the movie, then there is no way to say which direction the time is flowing and the entropy does not increase anymore.

This a brief account of what it is usually called the probabilistic interpretation of the second principle of thermodynamics, and it is not something very new by the moment that it was initially proposed by Boltzmann himself at the beginning of last century.

Is this the end of the story? Actually, not. There is a quite subtler aspect that is associated with the conservation of energy. It is known as Poincaré recurrence theorem. It states that in a system where energy is conserved, the dynamics evolve in such a way that, after a sufficiently long time, it returns to a state arbitrarily close to any given state. The time that we have to wait in order to observe this recurrence is called the *Poincaré recurrence time*. In simple words, this means that not only the dissipation of energy is an illusion because the energy is simply redistributed among all the dof, but also, that this redistribution is not final. If we wait long enough we will see that after some time, the energy will flow back to its initial distribution and our piston will get its potential energy back (with the gas particle becoming slightly colder) and the entropy increase is reversed. Thus, not only the decrease of entropy for a system composed by many dof is not impossible, but it will eventually happen. Nevertheless, what we have said before remains true: for *any finite observation time* the dynamic system evolves most probably in a direction where the entropy increases because this is the most probable evolution. However, if we wait long enough also the less probable outcome will be realized and thus the second principle violated.

How much time should we wait? How long is the Poincare recurrence time? The answer depends on the number N of our isolated (energy conserving) system. The larger N the exponentially longer is the time to wait.

4.3 The Switch Dynamics in the Presence of Fluctuations

Back to our main topic: the switch dynamics. Based on the previous discussion in Chaps. 2 and 3, we are now in position to express the dynamics of our logic switch in terms of a Langevin equation, like in Sect. 3.2, where the dissipative term is represented, for simplicity, by a simple viscous damping $-\gamma \dot{x}$ and the fluctuations by the stochastic force $\xi(t)$, with the confining potential V:

$$m\ddot{x} = -\gamma \dot{x} + \xi(t) - \frac{dV(x)}{dx} \tag{4.1}$$

where the confining potential is appropriately chosen for representing *combinational* or *sequential* switches. It is important to keep in mind that the switch dynamics $x(t)$ is represented by a stochastic process [2] whose time series is a member of a family of random functions, with given statistical properties. The most relevant statistical quantities are the average $\langle x \rangle$ and its variance $\sigma^2 = \langle x^2 \rangle - \langle x \rangle^2$ together with the probability density function $p(x)$, defined as:

$$Pr[a \leq x \leq b] = \int_a^b p(x)dx$$

where $Pr[a \leq x \leq b]$ represents the probability of finding x in the interval $[a \leq x \leq b]$.

If $V(x)$ is a quadratic potential, then $p(x)$ is a Gaussian function [3]. Clearly, if we have knowledge of $p(x)$ we can obtain all the other quantities. As an example:

$$\langle x \rangle = \int_{-\infty}^{+\infty} xp(x)dx$$

According to the thermodynamic (large N) picture discussed in the previous paragraph and, more extensively, in the previous chapter, we need to identify the microstate representing our system. An interesting step in this direction has been developed in recent times, under the name of stochastic thermodynamics [4]. In this framework, the microstates are identified by the value assumed by $x(t)$, i.e. the stochastic trajectory of our relevant observable.

At difference with the traditional Boltzmann approach, due to the presence of the confining potential $V(x)$, not all the values of $x(t)$ have the same probability to be realized. Thus, in order to introduce a proper entropy, we should adopt W. Gibbs definition of thermodynamic entropy for microstates, identified by the index i and characterized by different probabilities p_i:

$$S = -k_B \sum_i p_i ln(p_i)$$

The available volume in the phase space is measured by the number of available microstates; thus, the sum is taken over all possible microstates, and each of them is weighted by the probability that our system actually realizes such a microstate.

The extension to a continuum of microstates was proposed by C. Shannon (1916–2001), few years later [5], as:

$$S = -k_B \int_{-\infty}^{+\infty} p(x) ln(p(x)) dx$$

This form of the Shannon entropy is sometimes called *differential entropy* in the literature [6], and it should be handled with some care, because has some peculiar mathematical properties that differ from the Gibbs entropy.

By the moment that the probability of finding our switch variable $x(t)$ in the interval $[-\infty, +\infty]$ is clearly one, the spread of the probability density provides us an intuitive understanding of the entropy associated with different probability densities. As an example, in Fig. 4.1 we present two different arrangements for the probability density associated with a harmonic potential in the presence of Gaussian noise, at equilibrium. The temperature T, representative of the intensity of the fluctuations, is the same for the two arrangements, and the area below the probability density is one in both cases. However, the value of the potential $V(x)$ in Eq. (4.1) is different, being shallower in the left picture compared to the right one. The shallower is the potential, the spreader is the probability density, and the bigger is the associated entropy.

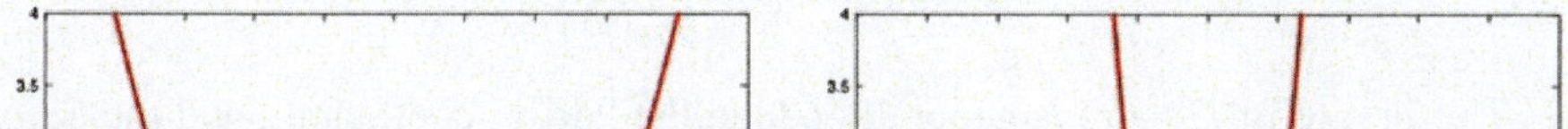

Fig. 4.1 Probability density p(x) (in blue) for two different values of the harmonic potential. The potential in the right picture is stiffer than in the left one (by a factor 3). As a consequence, the probability density is more confined, and the entropy is smaller

By the moment that $x(t)$ is a stochastic process and any actual realization is a random function, the usual definition of switch event from S_0 to S_1, based on $x(t) > x_{TH}$, does not hold anymore. We need a new definition that considers the fluctuating character of $x(t)$. A possible solution is the introduction of the notion of probability of a given state. We may indicate with p_0 and p_1 the probability of our switch to be in state S_0 and state S_1, respectively. With:

$$p_0 = \int_{-\infty}^{x_{TH}} p(x)dx$$

$$p_1 = \int_{x_{TH}}^{+\infty} p(x)dx$$

With $p_0 + p_1 = 1$.

Under this perspective, we can define a switch event by comparing the different probabilities. We say that the system is in the state S_0 with probability p_0 and in state S_1 with probability p_1. If $p_0 > p_1$ we will consider the switch from S_0 to S_1 and vice versa if $p_1 > p_0$ we will consider the switch from S_1 to S_0. In this perspective, the switch event is associated with the change of the probability density $p(x)$. In general, $p(x)$ is a function of time, thus we will write $p(x, t)$. Its evolution can be monitored by computing it from the statistics of $x(t)$ or it can be described with a partial derivative differential equation, known as Fokker–Plank equation[7], after Adriaan Fokker (1887–1972) and Max Planck (1858–1947) that proposed it in 1914 and 1917 respectively, as a general diffusion equation. The equation that specifically describes $p(x,t)$, in our case, is sometimes associated with the name of Marian Smoluchowski (1872–1917) that proposed it in 1916, as a possible description of the probability density evolution for a particle subjected to Brownian motion. We have:

$$\frac{\partial}{\partial t} p(x, t) = \left(-\frac{\partial}{\partial x} D^{(1)}(x) + \frac{\partial^2}{\partial x^2} D^{(2)}(x) \right) p(x, t)$$

The first term $D^{(1)}(x)$ is generally called the "drift" coefficient, and the second term $D^{(2)}(x)$ is called the "diffusion" coefficient. For a free particle, the drift coefficient is null, while for a particle that has no fluctuations the diffusion coefficient is null.

In our case, we are faced with a constant diffusion coefficient (equilibrium thermal bath) and a drift term provided by the force associated with the confinement potential of the switch.

We have:

$$\frac{\partial}{\partial t} p(x, t) = \frac{\partial}{\partial x} \left(\frac{\partial V(x)}{\partial x} p(x, t) \right) + T \frac{\partial^2}{\partial x^2} p(x, t)$$

where $V(x)$ is the potential and T the temperature of the thermal bath. In general, the Fokker–Plank equation is a second-order partial differential equation, and in order to find its solutions, we need an initial condition $p(x, t_0)$ and boundary conditions for the space variable. Analytical solutions can be found in a limited number of cases. Among these there is the harmonic potential case, where the stationary solution is the already mentioned Gaussian probability density.

4.4 Combinational Switches: The Switch in the Presence of Fluctuations

Back in Chap. 2, at Sect. 2.8 we discussed the switch event for a deterministic switch. The same reasoning can be applied here. In this case, however, instead of focusing on the representative point particle, we will focus on the evolution of the probability density. In Fig. 4.2, we present pictorially an example of the evolution of $p(x, t)$ in the presence of a switch from the state S_0 to the state S_1. In both cases, i.e. before and after the switch, due to the statistical character of the event itself, we will always have a finite probability to find the system in the unwanted state. This condition gives place to the switch error phenomena. We will deal more with this aspect in the next paragraphs.

As before, the switch event S_0 to S_1 can be generated by applying a deterministic constant force F_{ext} whose amplitude is properly chosen in order to satisfy the condition $p_1 > p_0$, for a given threshold x_{TH}. The switch event S_1 to S_0 is obtained once the force F_{ext} is removed.

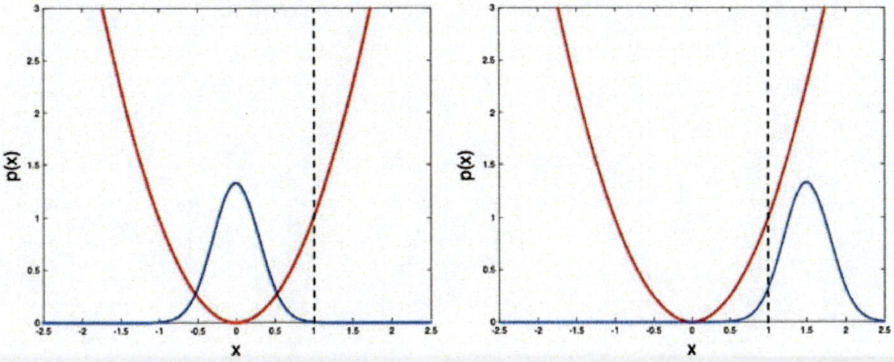

Fig. 4.2 Representation of the two states for a combinational switch. The threshold x_{TH} is marked by the vertical dashed line. Left: the system is the state S_0, characterized by $p_0 > p_1$. Right: the system is the state S_1, characterized by $p_1 > p_0$

4.5 Sequential Switches: The Switch in the Presence of Fluctuations

For sequential switches, in the presence of fluctuations, the two states S_0 and S_1 can be represented according to Fig. 4.3.

At difference with the case of combinational switches, we have to note that for the sequential switches, the states S_0 and S_1 are not stationary states. As a matter of fact, the stationary solution of the Fokker–Plank equation in this case is the one represented in Fig. 4.4 where, due to the symmetry of the potential, there has to be the same probability to be in state S_0 and S_1.

The fact that the two states S_0 and S_1 are not stationary has important consequences of the use of the sequential switches. It is interesting to study the time evolution of $p(x,t)$ as a solution of the Fokker–Plank equation for this "double well potential". If we set the initial condition $p(x, t_0)$ as a Gaussian distribution with a given standard deviation σ_0 and boundary condition such that $p(\infty, t) = 0$ (due to the confining character of the potential), the time evolution of $p(x, t)$ is represented in Fig. 4.5.

In the first panel (a), at time $t_0 = 0$ we set the initial probability density with the system in state S_0. As time passes (b), the probability density evolves towards a local state of equilibrium, with the standard deviation σ that grows from σ_0 to σ_w (c). The state of local equilibrium is reached with a characteristic time (local relaxation time) $t = t_w$. After further time has passed (d), a small peak of the probability density starts to appear also in the right well (e) and grows until it reaches the condition of global equilibrium state (f) with a symmetric $p(x, t)$. The global equilibrium is reached with a characteristic time $t = t_k$. Usually $t_k \gg t_w$.

Based on what we have seen in the previous chapter, this time evolution towards a global equilibrium state corresponds to the spontaneous transformation characterized by the tendency of the system entropy to reach its maximum.

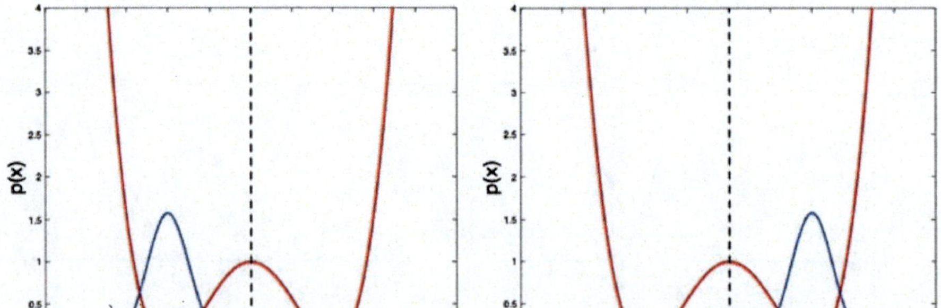

Fig. 4.3 Representation of the two states for a sequential switch. By the moment that the potential is symmetric, the threshold x_{TH} is set at $x = 0$. Left: the system is the state S_0, characterized by $p_0 > p_1$. Right: the system is the state S_1, characterized by $p_1 > p_0$

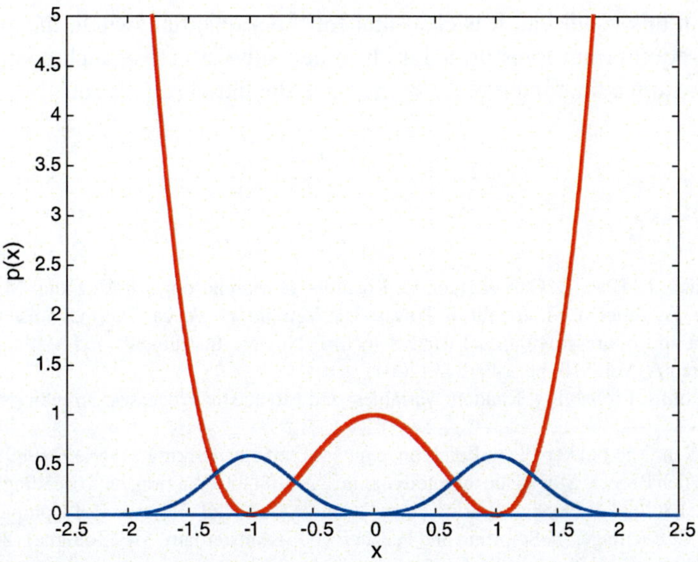

Fig. 4.4 Representation of the stationary solution, for a sequential switch

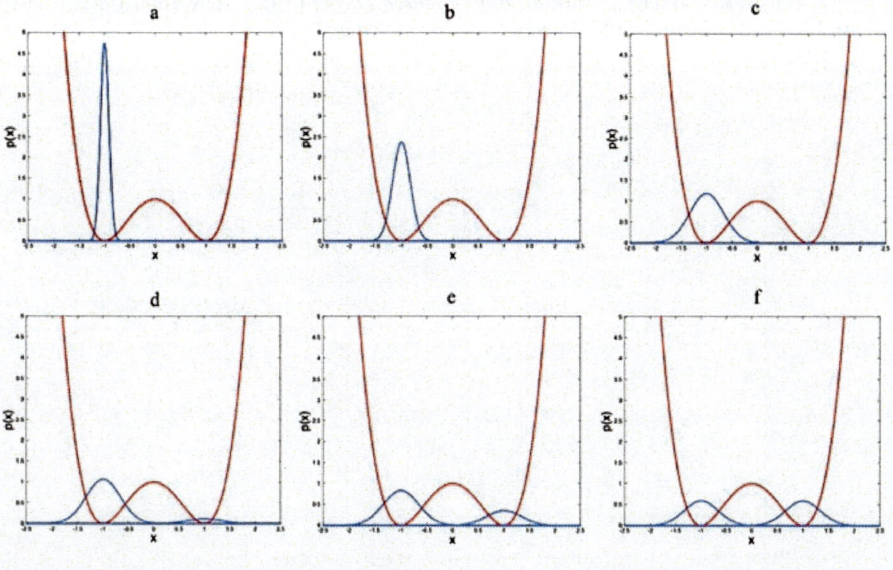

Fig. 4.5 Representation of the time evolution (from **a–f**) of the probability density for the initial state S_0. **a** At time $t_0 = 0$, the initial probability density has $\sigma = \sigma_0$. **b** As time passes, the probability density evolves towards a local state of equilibrium and σ grows. **c** A local (inside one well) state of equilibrium is reached with $\sigma = \sigma_w$. **d** After further time has passed, a small peak of the probability density starts to appear also in the right well. **e** It grows until it reaches the condition of global equilibrium. **f** Global equilibrium state is reached

Within this condition, it is clear that for the sequential switch, the switch event is a non-equilibrium transformation. In order to be effective, such a switch should happen within a time interval $\Delta t \ll t_k$, since the initial state is set.

References

1. P. Amico, L. Bosi, L. Gammaitoni, G. Losurdo, F. Marchesoni, M. Mazzoni, M. Punturo, R. Stanga, A. Toncelli, M. Tonelli, F. Travasso, F. Vetrano, H. Vocca, Thermal noise reduction for present and future gravitational wave detectors. Nuclear Instruments and Methods in Physics Research A, Vol. 518, pp. 240–243 (2004)
2. A. Papoulis, Probability, Random Variables, and Stochastic Processes, 3rd edn (McGraw-Hill, 1991)
3. H. Risken, The Fokker-Plank Equation, cap. 5.3, 2nd edn. (Springer-Verlag, 1989)
4. C. Van den Broeck, Stochastic thermodynamics: A brief introduction, in Proceedings of the International School of Physics "Enrico Fermi",Course CLXXXIV "Physics of Complex Colloids", ed. by C. Bechinger, F. Sciortino and P. Ziherl (IOS, Amsterdam; SIF, Bologna) (2013). https://doi.org/10.3254/978-1-61499-278-3-155
5. C. Shannon, A mathematical theory of communication. The Bell System Technical Journal **27**, 3 (1948)
6. M. Thomas, J.A. Thomas, *Elements of Information Theory* (Wiley, New York, 1991)
7. H. Risken, The Fokker-Plank Equation (Springer-Verlag, 1989)

Chapter 5
Fundamental Energy Limits in Computation

I learned early that you only have so much energy to give. You have to spend it correctly.
Eva Gabor

Abstract In the previous chapter, we have discussed the physics framework for describing the dynamics of binary switches, the fundamental bricks in the computer building, and we are now in position to address the question of how much energy does it require to perform a computation. There are interesting surprises in store.

5.1 How Much Energy Does It Take to Switch and Why This is Important?

As we have anticipated, by the moment that all the computing tasks are executed via logic gates and their functioning is performed by assembling binary switches, we should focus on the energy budget required by a single switch. In the following, we will discuss the combinational switch and the sequential switch separately. In the next chapter, we will study, instead, the fundamental limits in energy dissipation of memories.

The questioning of the minimum amount of energy required for performing computation, far from being a mere theoretical curiosity, has profound implications in the practical realization of future generations of computing devices. In turn, this has relevant societal and economic implications. Computing devices are the cornerstone of a whole technology sector called Information and Communication Technology (ICT) that plays a strategic role in the world economy. Its impact on cultural and social development is presently enormous, and it will keep growing in the foreseeable future. What is important to consider, however, is that state of the art ICT is based on computing devices whose functioning is currently dominated by power dissipated in heat. The overall amount of energy dissipated during computation, according to the SMART 2020 [1] report, is close to 5% of the worldwide energy consumption. *Given that the use of ICT will further increase and the overall energy consumption*

L. Gammaitoni, *The Physics of Computing*,
https://doi.org/10.1007/978-3-030-87108-6_5

will hopefully decrease due to the help of ICT and other measures, it is expected that the share of ICT on the worldwide energy consumption will grow in the future. Carbon dioxide emissions from the use of ICT are therefore presently increasing. Hence, it becomes more and more important to consider and improve the energy efficiency of ICT. On the short term, it will be an obvious and practical solution to better exploit the potential of technologies that already exist or are currently in the making. On the long term, new and disruptive ideas will be needed [2].

It appears clear that, in order to design, make and operate the "new and disruptive ideas" for computing technologies, a deep understanding of what is the minimum required amount of energy per each computation step is necessarily required.

5.2 Combinational Switches: Minimum Dissipated Energy

In the previous chapter, we have discussed the switch event for the combinational switch (see Fig. 4.2).

We note that states S_0 and S_1 do not have the same energy due to the shape of the potential $V(x)$. If we indicate with $\Delta E = V_1 - V_0$ the potential energy difference between the two states, it is clear that, in order to perform the switch, the deterministic force F_{ext} has to perform a work at least equal to ΔE, against the potential energy. What about the other forces in the Langevin Eq. (4.1)? The stochastic force $\xi(t)$ has zero mean; thus, on average, the work performed is zero. Finally, the dissipative force $\gamma\dot{x}$ performs a work that is proportional to the speed of the switch. In the adiabatic limit, i.e. when the switch event is produced with $\dot{x}\rightarrow 0$, the dissipated energy tends to zero.

Another more elegant way to express the contribution of the dissipative and fluctuation forces is to consider, according to stochastic thermodynamics, the heat exchanged Q, by averaging over many stochastic trajectories the Stratonovich integral [3].

$$Q = \left\langle \int_0^t - \frac{dV_e(x)}{dx} dx - \int_0^t m\dot{x}d\dot{x} \right\rangle$$

where $V_e(x) = V(x) - F_{ext}$ is the effective potential, already introduced in Chap. 2. Also in this case, by extending the integration interval, i.e. by proceeding arbitrarily slowly, the heat exchanged can be made arbitrarily small.

On the other hand, if we adopt the pure thermodynamics point of view, we have to consider the existence of a thermal bath that exchanges a certain amount of heat equal to $Q = T\Delta S$. However, due to the linearity of the potential, with a constant F_{ext}, the system entropy at the beginning and at the end of the switch is the same. Thus, during a reversible (i.e. quasi-equilibrium transformation) we have $\Delta S = 0$ and no minimum heat is exchanged.

Moreover, the work done by the deterministic force F_{ext} during the S_0 towards S_1 switch is the opposite of the work done during the S_1 towards S_0 switch. Based on this analysis, if we are capable of performing the switch event with arbitrarily small velocity, the minimum required energy expenditure is ΔE. Such energy can be completely recovered during the switch back S_1 to S_0. These events, being on average equally likely in a long computation task, let us conclude that it is possible, at least in principle, to perform such a task by dissipating zero energy.

In order to better illustrate this point, we performed an experiment [4] where a microelectromechanical cantilever has been employed as a combinational switch (Fig. 5.1). The device consists of a single combinational switch made with a V-shaped Si_3N_4 cantilever, 200 μm long, that can be bent by applying electrostatic forces with an electrical probe placed close to the cantilever tip. Once the electrostatic force is applied, the position of the cantilever is changed. The force is applied through an electrostatic probe consisting of a tungsten tip (100 nm tip radius) placed at 5 μm from the cantilever tip.

In Fig. 5.2, we show the measurement apparatus for the tip position. The cantilever position is measured, under vacuum, with an optical lever: the deflection of the laser beam (633 nm) due to the bend of the cantilever is detected by a two-quadrant photo detector.

In Fig. 5.3, we show the results of the measurement of the tip position with and without the electrostatic force applied. The presence of fluctuations in the tip position is well apparent by the Gaussian-like distribution of the data. The vertical dashed line in the figure represents the threshold between the two states S_0 and S_1.

The scope of the experiment was to estimate the amount of energy dissipated during a cycle of two subsequent switches S_0 to S_1 and S_1 to S_0; thus, we measured the work L performed on the system by the external forces during this entire operation cycle. Since, in a cycle the change in the internal energy U of the system is zero, we have Q = L. In order to measure the work performed during the cycle, we computed the Stratonovich integral between $t = 0$ and $t = \tau_p$, being τ_p the cycle duration. The

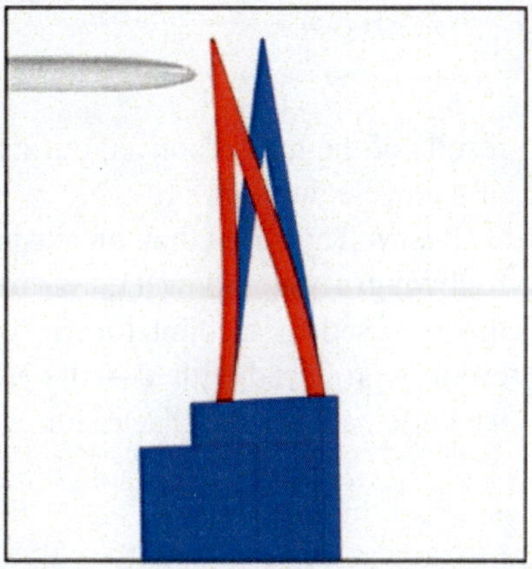

Fig. 5.1 Pictorial representation of the cantilever. When a conductive probe is charged, an electrostatic force is applied to the cantilever that bends towards the probe (red)

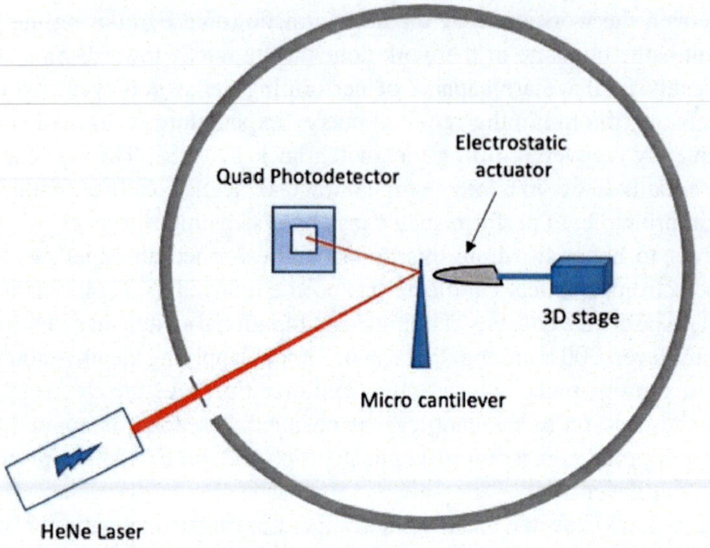

Fig. 5.2 Scheme of the set-up used for the measurements. The red line is a representation of the optical path of the laser used for the measurements

Fig. 5.3 Statistical distribution of the cantilever tip position without (left) and with (right) the electrostatic force. Position is expressed in nm. The vertical dashed line represents the threshold between S_0 and S_1 states

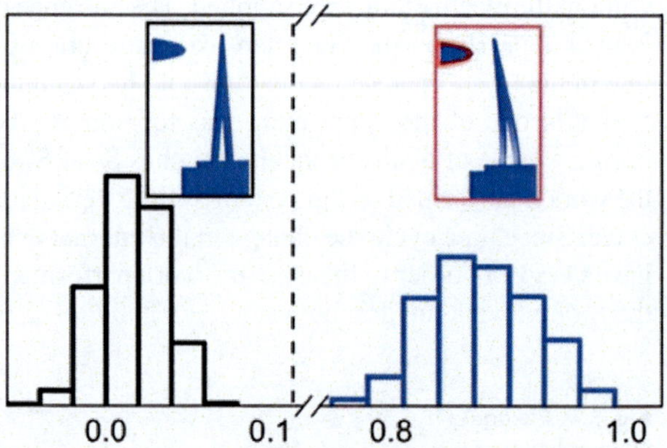

results of the heat produced during the cycle are presented in Fig. 5.4 as a function of the cycle duration τ_p.

It is well apparent that, increasing the cycle duration, the produced heat decreases following a power law (τ_p^{-3}), in good agreement with a friction model (Zener theory) used to account for the velocity dependent dissipation. When the switch event is produced with $\dot{x} \rightarrow 0$, i.e. $\tau_p \rightarrow \infty$ the dissipated energy tends to zero, as we anticipated in our discussion.

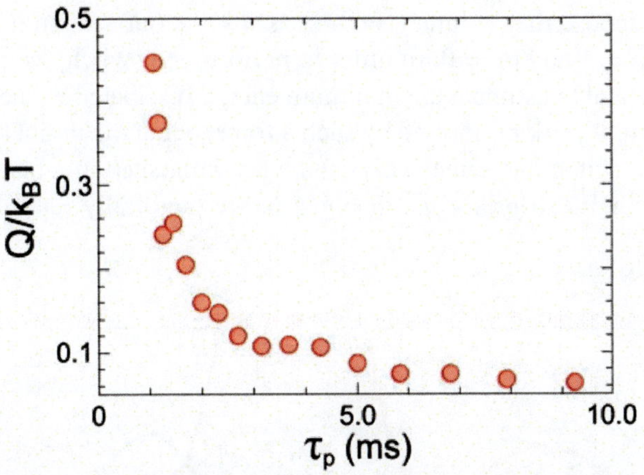

Fig. 5.4 Average heat production during the cycled combinational switch operation as a function of cycle time τ_p. By increasing the cycle time, the produced heat decreases following a power law, in agreement with the dissipative model considered here. The average heat Q is shown here in $k_B T$ units, with T being the room temperature

5.3 Sequential Switches: Minimum Dissipated Energy

A reasoning similar to the one developed in the previous paragraph can also be applied to the sequential switch event of Fig. 4.3 (reproduced here as Fig. 5.5, for convenience). However, in this case we need to remember that the time evolution of the switch has to consider that the initial and final states are not equilibrium states. As we concluded in Sect. 4.5, if we call t_k the global relaxation time, the switch event should happen within a time interval $\Delta t \ll t_k$, since the initial state is set. Due to the symmetry of the two states, the switch S_0 to S_1 and the switch S_1 to S_0 can be treated on the same base and, for sake of simplicity, we will focus our attention only on S_0 to S_1.

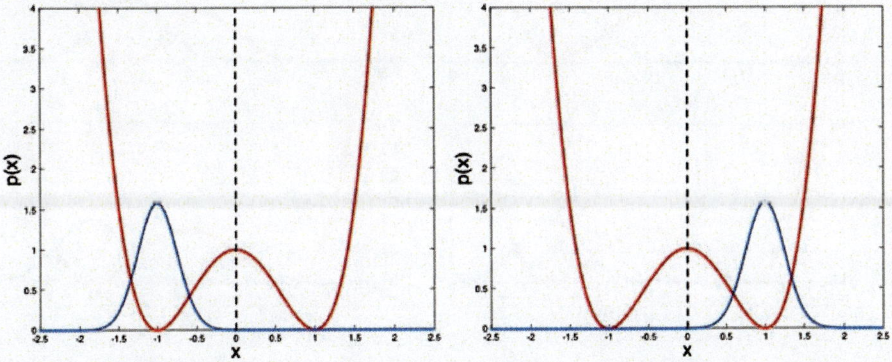

Fig. 5.5 Representation of the two states for a sequential switch. Left: the system is the state S_0, characterized by $p_0 > p_1$. Right: the system is the state S_1, characterized by $p_1 > p_0$

So, let's suppose that at time t, with $t_w \ll t \ll t_k$ our system is in state S_0 with probability $p_0 = 1$ and $p_1 = 0$. In order to perform the switch, we need to apply an external force and to estimate the minimum energy dissipated we need, as before, to compute the total work performed by such a force, being the initial and final internal energy of our switch, the same. This can be accomplished in a number of different ways that we will call *protocols*. In Fig. 5.6, we pictorially represent two of such *protocols*.

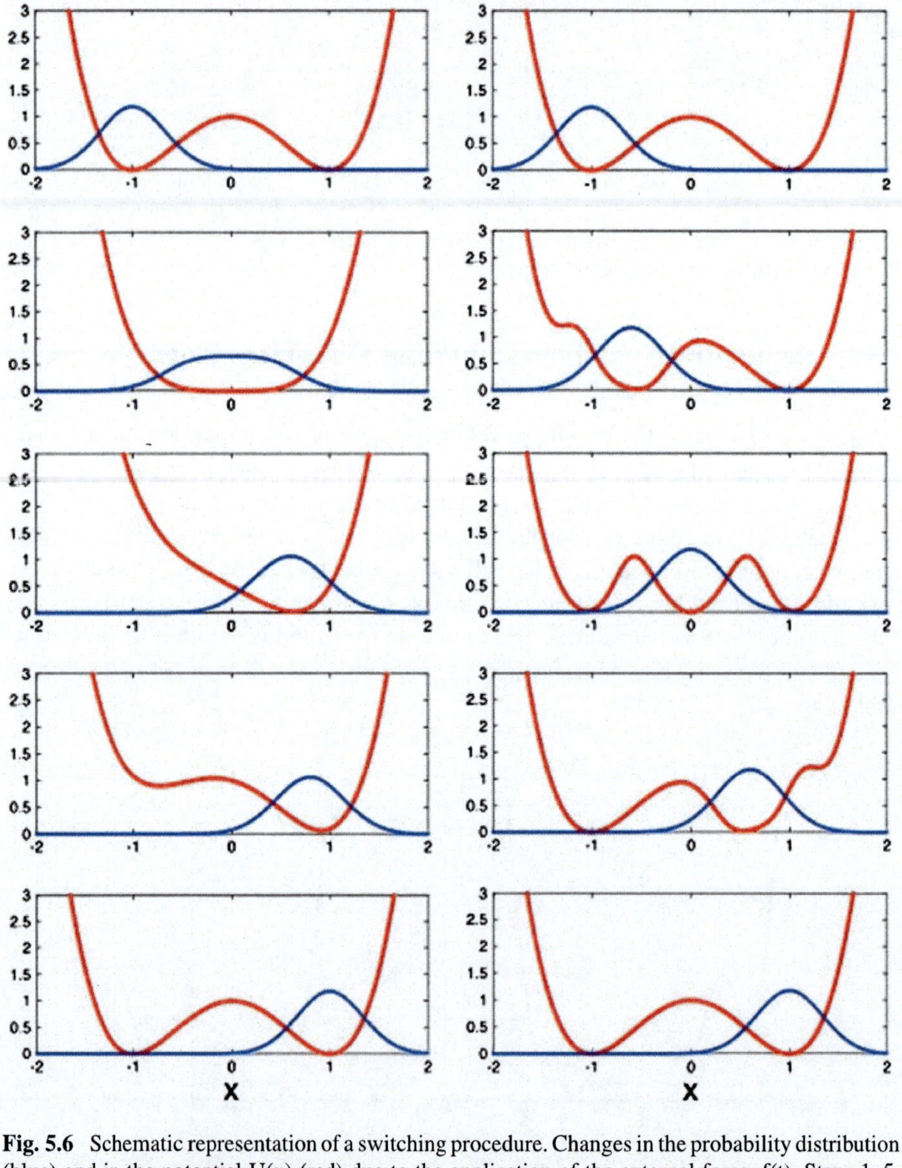

Fig. 5.6 Schematic representation of a switching procedure. Changes in the probability distribution (blue) and in the potential U(x) (red) due to the application of the external force, f(t). Steps 1–5, from top to bottom. Left: barrier drop *protocol*. Right: zero-power *protocol*.

The first *protocol* (left column in Fig. 5.6) is based on the sudden drop of the potential barrier between the two states and the subsequent tilt of the potential towards the destination state S_1. This action produces a rapid increase of the width of the probability distribution with the associated increase of the system entropy. When the tilt of the potential is applied and the barrier is restored, a necessary entropy reduction is required. In this case, unless the barrier drop is performed very slowly, the sudden increase of the entropy happens under a far-from-equilibrium condition, similar to the free expansion of a perfect gas. While this increase can be performed at no work expenses, the subsequent entropy reduction requires some heat to be produced (work to be performed). On the other hand, the second *protocol* (in the literature often indicated as "zero-power *protocol*" [5]) is based on the application of an external force that does not change the probability distribution throughout the entire duration of the *protocol*. The probability density here is "simply" shifted from the state S_0 (below $x = 0$ threshold) to the state S_1 (above $x = 0$ threshold). The main advantage of this *protocol* is that the entropy never changes during the switch and thus no external work is actually required.

As for the combinational switch, also in this case, if the proper conditions are met, the switch operation can be conducted by dissipating zero energy.

5.4 Bit Reset: Minimum Dissipated Energy

In Sect. 4.5, we have discussed the time evolution of the probability density for the sequential switch (Fig. 4.4). We have seen that, once the initial system state is set (at S_0 or S_1), the local equilibrium is reached within a characteristic time (local relaxation time) $t = t_w$. After some further time has passed, the dynamics of x(t) starts to be characterized by oscillations around the potential minima, with occasional random crossings of the potential barrier separating the two minima. If we wait long enough, we can observe that after a time, t_k, it will jump into the right well and eventually back into the left well, and so on at a constant pace and a global equilibrium condition is reached when the probability distribution symmetrically divided between the two states. t_w and t_k are random variables whose average values are usually referred to as *intrawell* relaxation time and *interwell* relaxation time, respectively. They represent the average time the system takes to establish local equilibrium within one well (as it would be if the potential was no wider than a single well) and the average time it takes to go to global equilibrium in the entire potential, respectively. By the moment that t_k depends exponentially on the barrier height, in practical switches the barrier height is chosen to be large enough to guarantee $t_k \gg t_w$.

It is important to stress, once again, that this relaxation towards global equilibrium process is a spontaneous transformation where the entropy of the system naturally increases.

In the description of the switch event for the sequential switch, we assumed that the system is at local equilibrium in a given well and that the switching process proceeds within a time, Δt, such that $\Delta t \ll t_k$. Now, let's suppose that instead we

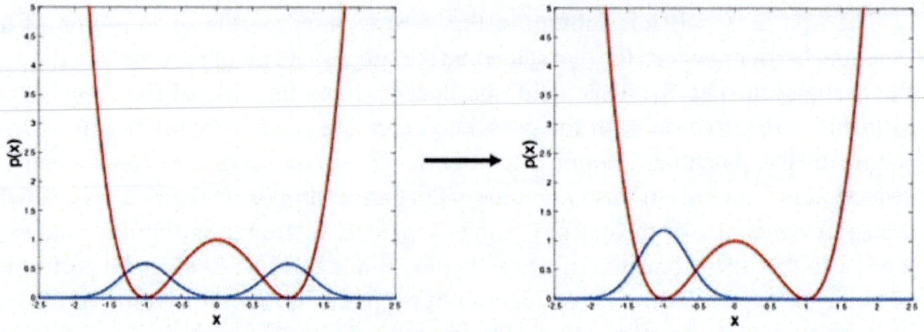

Fig. 5.7 Schematic representation of a reset procedure

are not at local equilibrium anymore, but rather have reached a global equilibrium, i.e. before starting the switch process, we have waited a time, $\Delta t > t_k$. As we have seen, we are now with a situation where the probabilities to be in state S_0 and S_1 are the same, i.e. $p_0 = p_1 = ½$.

If we want to perform a switch operation, say S_0 to S_1, we need to operate a transformation that changes the probability density distribution in order to reach the condition $p_0 = 1$, $p_1 = 0$. This transformation is usually called *reset* (Fig. 5.7). From this perspective, we can say that a switch operation transforms a local equilibrium condition into another local equilibrium condition, while a reset operation transforms a global equilibrium condition into a local equilibrium condition.

Similarly to what we have done for the switch operations, we can identify a proper procedure that minimizes the energy required to operate the reset by zeroing the work done by the deterministic forces. In this case, however, the resulting change in entropy cannot be made null because the initial (before the reset) and final (after the reset) entropies are not the same, and the minimum amount of work to be done is $L = Q = T\,\Delta S$.

The change in entropy ΔS depends on the initial and final probability densities, i.e. the probability density at time t_i and t_f:

$$\Delta S = -k_B\left(\int_{-\infty}^{+\infty} p(x,t_f)ln\big(p(x,t_f)\big)dx - \int_{-\infty}^{+\infty} p(x,t_i)ln(p(x,t_i))dx\right) \quad (5.1)$$

In the simple case where we have two microstates coincident with S_0 and S_1, with $p_{0i} = p_{1i} = ½$ and $p_{0f} = 1$ and $p_{1f} = 0$ (reset to S_0 operation), we can use Boltzmann expression to compute the change in entropy:

$$S = k_B ln W$$

With *W* the number of possible configurations. In the initial state, we have two possible configurations (two possibly occupied microstates), while in the final state we have just one configuration (1 possibly occupied microstate). Thus

$$\Delta S = k_B(ln1 - ln2) = -k_B ln2$$

As we anticipated, by the moment that the reset operation requires a decrease of the entropy, there is a minimum energy to be dissipated equal to

$$\mathrm{Q} = -k_B T ln2$$

Such a result is sometimes called Landauer's limit and, as we have shown, it is a simple consequence of the second law of thermodynamics. Ralph Landauer (1927–1999) developed this result in the framework of information theory, with much wider implications [6]. We will come back to these aspects in Chap. 7. For the moment we observe that, and difference with the switch operation, the reset operation cannot be accomplished at zero energy expenditure and it requires a minimum energy. How big is this minimum energy? That depends on the initial and final probability densities in our reset operation according to (5.1).

It is interesting to observe that, if we admit a certain probability of error in the reset operation, we can lower the Landauer's limit [7]. Let's suppose that during the reset to S_0 operation, an error occurs and we end up in the S_1 state, with a certain probability, $P_e = 1 - p_0 = p_1$. In Fig. 5.8, we present the probability densities before and after the reset to S_0 with non-null error probability.

To compute the minimum energy for this reset operation, let's proceed with the computation of the entropy change. We will use the Shannon entropy formulation:

$$S = -k_B \int_{-\infty}^{+\infty} p(x) ln(p(x)) dx$$

The entropy before the reset ($p_0 = p_1 = ½$) is promptly computed as $S_i = k_B ln\ 2$. After the reset operation, we have:

$$S_f(P_e) = -k_B((1 - P_e)ln(1 - P_e) + P_e ln(P_e))$$

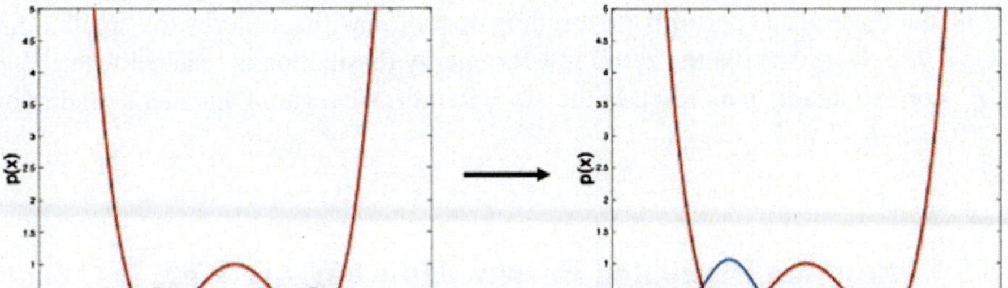

Fig. 5.8 Schematic representation of a reset procedure with non-null error probability

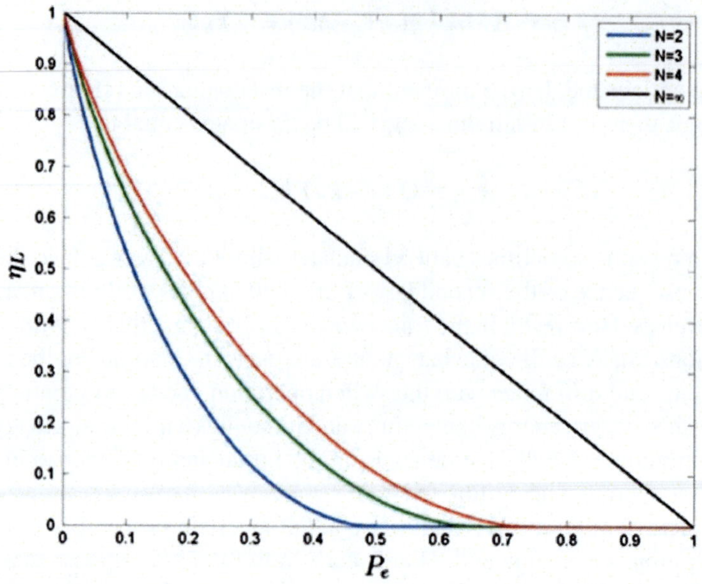

Fig. 5.9 Energy ratio ηL as a function of the error probability, P_e. It is interesting to note that for a bistable sequential switch (N = 2), if we accept an error probability of 20% ($P_e = 0.2$), we need to dissipate approximately 1/4 of the minimum required by the Landauer limit. Multi-stable switches (N = 3, N = 4) show an energy ratio, ηL, larger than the bistable one. In the large N limit, ηL, becomes a linear function of P_e

Expressed as a function of P_e, with $p_0 = 1 - P_e$. As a consequence, the minimum energy to be dissipated during the reset operation is a function of the error probability:

$$Q(P_e) = -k_B T((1 - P_e)\log(1 - P_e) + P_e ln(P_e) - ln2)$$

In Fig. 5.9, we plot the energy ratio $\eta L = Q(P_e)/(k_B\ T\ ln\ 2)$ as a function of the error probability, P_e. As one can see, for $P_e > 1$, we have $\eta < 1$, implying that if we are willing to accept a larger-than-zero error probability, we can beat the Landauer's limit and perform the resetting operation with an energy toll smaller than k_B T ln 2. Consistently, the zero limit for energy dissipation is reached when $P_e = 1/2$, corresponding to no reset. In the $P_e = 0$ limit, we regain Landauer's prediction, $Q(0) = -k_B\ T\ ln\ 2$.

5.5 Minimum Dissipated Energy: How Far Are We?

As we have seen in the previous paragraph, provided that we operate the logic switches according to certain protocols, we can at least in principle perform a switch operation by spending zero energy. This is an important result by a fundamental point of view and has some interesting implications. Present-day computers are quite complex systems, where computation is made with a large number of logic gates composed in turn by logic switches. If we exclude the reset operation, that could be performed once for all and taking proper care of the relaxation processes, the results that we obtained before implicates that the whole computation process could be, at least in principle, executed without spending any energy.

How far are we from this goal? The computer industry for many decades has focused on the scaling of existing technology towards smaller dimensions, in order to achieve high-speed and high-performance computing. Power dissipation was not at the centre of attention until the year 2000 [8]. At this point, the critical dimensions in complementary metal-oxide semiconductor (CMOS) devices, specifically field-effect transistors (FETs), became sufficiently small that the static power dissipation through leakage and parasitic currents became larger than the dynamic power dissipation from switching [9]. Large dissipated energy implies high operating temperature with potential risks for the device survival.

The energy dissipated by a FET during a switch event can be expressed as:

$$E_{MOSFET} = \alpha C_L V_{DD} f + I V_{DD} \tau + P_{SC} \tau$$

where α is the activity factor, C_L is the load capacitance, f is the circuit clock frequency, V_{DD} is the supply voltage, τ is the switching time (roughly the inverse of the clock frequency), I is the leakage current and P_{SC} is the short-circuit power dissipation.

Reducing the supply voltage V_{DD} has been for some time the main strategy in trying to reduce the dissipated energy; however, the finite band-gap of a semiconductor provides a lower limit below which the transistor cannot switch. Also changing the semiconductor material in favour of a smaller band-gap is not a very successful strategy because the minimum supply voltage is a function of the so-called *subthreshold slope*. This is a parameter that represents the slope of the device between the on-current (the logical "1") and the off-current (the logical "0"). For optimized devices, the minimum *subthreshold slope* only depends on fundamental constants and temperature. For a device at 300 K, the minimum slope is about 60 mV/decade which fixes a minimum threshold voltage.

In Fig. 5.10, we present a comparison for existing and designed switches, together with the figure for an average biological neuron and synapse. As it is well apparent, there is still much room for improvement.

The data were collected during the preparation of the Strategic Research Agenda, an official document of the ICT-Energy Coordination Action, funded by the European Commission (G.A. N. 611004).

The legend at the right: Key: CMOS HP (high-performance CMOS), CMOS LP (low-power CMOS), iii-vTFET (III-V tunnel FET), HJFET (heterojunction FET), gnrFET (graphene nanoribbon TFET), GpnJ (graphene pn-junction), spinFET (Sughara-Tanaka spin FET), SST/DW (spin torque domain wall), STMG (spin torque majority device), STTtriad (spin torque triad), STO (spin torque oscillator), ASLD (all spin logic device) and NML (nanomagnetic logic).

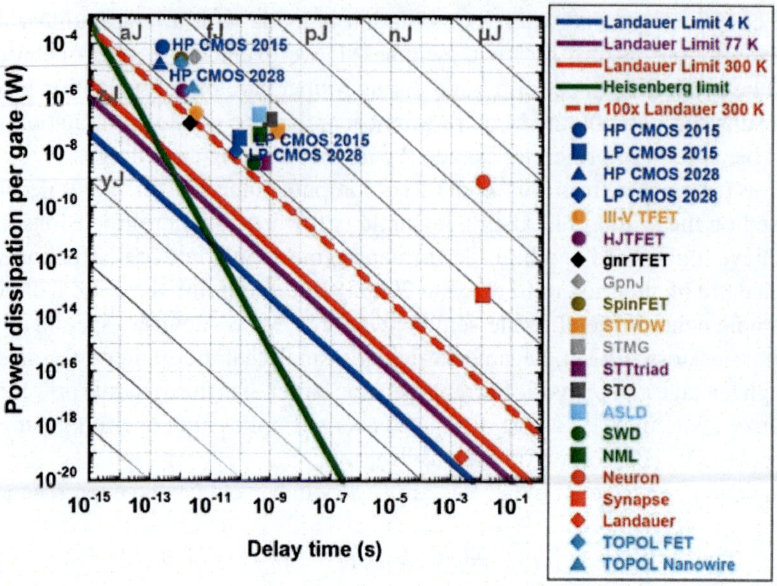

Fig. 5.10 Intrinsic power dissipation per device switch versus delay time for CMOS and future devices as proposed in the ITRS future emerging technology roadmap 2013. The diagonal grey lines are lines of constant energy. The figure indicates a limit of about 1 aJ per switch for CMOS devices and 250 zJ per device for the best proposed future switching device

Switches represented by research prototypes are indicated as: Landauer (red diamond) and is the result obtained during the EC funded Landauer Project, whose details are discussed in Sect. 7.4. The Si CMOS FET and the Si nanowire from the EC-funded project TOPOL.

References

1. "SMART 2020: enabling the low carbon economy in the information age" is a report published by "The Climate Group", an independent, not-for-profit organization. http://www.theclimategroup.org/_assets/files/Smart2020Report.pdf
2. Disruptive Solutions for Energy Efficient ICT, Expert consultation, FET Proactive, 2010. European Commission
3. U. Seifert, Stochastic thermodynamics, fluctuation theorems and molecular machines. Rep. Prog. Phys. **75**, 126001 (2012)
4. M. López-Suárez, I. Neri, L. Gammaitoni, Sub-kBT micro-electromechanical irreversible logic gate. Nat. Commun. **7**, 12068 (2016)
5. L. Gammaitoni, D. Chiuchiú, M. Madami, G. Carlotti, Towards zero-power ICT. Nanotech 2015, 26, 222001
6. R. Landauer, Dissipation and heat generation in the computing process. IBM J. Res. Dev. **5**, 183–191 (1961)
7. L. Gammaitoni, Nanoenerg. Lett. **5**, 10 (2013)

8. Strategic Research Agenda, ICT-Energy, EC 611004 (2011)
9. V. Zhirnov, R. Cavin, L. Gammaitoni, Minimum energy of computing, fundamental considerations, in ICT-Energy-Concepts Towards Zero-Power Information and Communication Technology, IntechOpen (2014)

Chapter 6
Storing Data

He knew the forms of the clouds in the southern sky on the morning of April 30, 1882, and he could compare them in his memory with the veins in the marbled binding of a book he had seen only once.
Jorge Luis Borges, Funes el Memorioso

Abstract An important part of the computing activity is storage and retrieval of the results of computation in special devices called memories. In this chapter, we discuss how a memory device works and what are the fundamental limits that physics poses to the functioning of memories. We will discover that, quite surprisingly, once written a memory can be maintained in time with an arbitrarily small amount of energy. However, no matter how much energy are we willing to invest, no memory can last forever.

6.1 Memories: How Do They Work?

Present generation of digital computers are based on an architecture that was initially proposed by John Von Neumann (1903–1957), Hungarian-born American scientist, in 1945. Von Neumann built on the computer scheme originally invented by the father of computers, Charles Babbage (1791–1871), that used to distinguish between the computing activity per sè and the storage activity: processed data are stored in special devices, called memories, in order to make them available for further use. Memories are nowadays important components of computers, and their functioning is of crucial importance for all the computing activities.

A digital memory is a device that is specifically designed to maintain a logic state over time. According to this aim, binary sequential switches (see Sect. 2.4) are the natural model devices to accomplish such a task. Moreover, by the moment that in digital devices the information is usually stored in the form of binary numbers, where the "0" bit is valued the same as the "1" bit, it is customary to consider a sequential switch with two equally energetic stable states separated by a potential barrier (see Fig. 6.1a).

L. Gammaitoni, *The Physics of Computing*,
https://doi.org/10.1007/978-3-030-87108-6_6

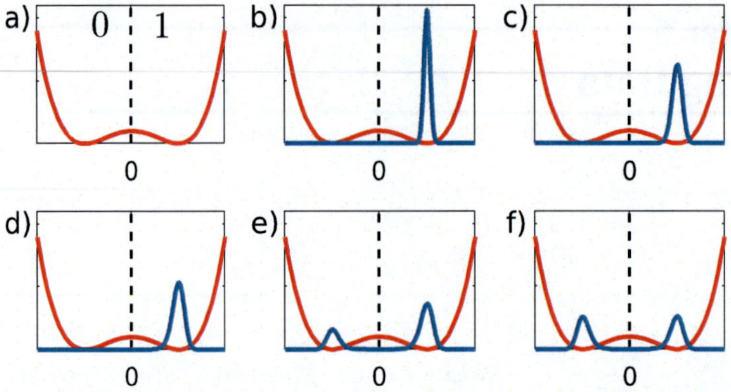

Fig. 6.1 **a** Sequential switch model for representing a digital memory. **b** A bit with logic value 1 is stored in this memory. According to the relaxation to equilibrium process, *p(x, t)* (blue curves) will first relax inside the right well (**c**) and (**d**) and then it will diffuse into the left well of the potential (**e**) to eventually reach global equilibrium (**f**). In this condition, the memory of the stored bit is definitely lost

When a memory is written, i.e. a logic bit "0" or "1" is encoded by selecting the state S_0 or S_1, the switch is set in a configuration similar to Fig. 6.1b, where, for sake of example, we have written bit "1" into the memory. If this operation is accomplished in a sequential switch, then a typical *Landauer reset* to "1" is performed.

Based on the discussion developed in the previous chapter, the physics of the device is well described by the time evolution of the probability density $p(x, t)$. The physical state corresponding to the bit "1", encoded into the sequential switch, corresponds to a non-equilibrium condition that evolves, within the system relaxation time τ_K, to thermal equilibrium (Fig. 6.1f). As this process takes place (Fig. 6.1c–f), memorization errors accumulate up to the point that the memorized bit is statistically lost when thermal equilibrium is achieved.

By the moment that it is our intent to memorize bit "1", it is clear that a perfect memorization condition is realized when $p_0 = 0$ and $p_1 = 1$. Let's suppose for a moment that it is possible to have an initial condition at time $t = 0$ that satisfies this requirement (we will see later that this is not the case). It is clear that, as time passes, due to the relaxation to equilibrium, a certain $p_0 > 0$ will start soon to appear. The existence of a $p_0 > 0$ can be interpreted as a probability of error in the bit memorization. In fact, the larger is p_0 the greater is the probability that our memory can be found to host a bit "0" instead of a bit "1", as initially desired.

In order to fight information loss, it is thus customary to set up a procedure that fights the relaxation process in the memory device. In modern digital memories, this is achieved through a cyclic *refresh operation*, consisting in a read-and-rewrite memory process, executed at periodic time intervals of duration t_R. The *refresh operation* requires a time t_p (protocol time) during which the memory is not available for use. In modern computing systems, t_p is few nanoseconds, t_R few milliseconds, while τ_K few seconds.

The relaxation to equilibrium process is a statistical one and, once the memory has been written, there is always a finite probability that within any given time $\bar{t}$ the system crosses the potential barrier and ends in the wrong logic state. If by chance, when the *refresh operation* takes place, the wrong bit is read, then the memory is overwritten with the wrong logic value. This occurs because the refresh operation does not correct errors. Thus, any memory preservation strategy is affected by a given probability of failure.

Specifically, we indicate with $P_E(\bar{t}, t_R)$ the memory loss probability, i.e. the probability to find the wrong value of the bit when the memory is interrogated during the entire time $\bar{t}$, since its first writing. In any practical application, it is interesting to set the acceptable probability P_E and the desired memory duration time $\bar{t}$, then to deduce what is the required t_R to be used for the refreshing period that satisfies those conditions.

6.2 Memory Dynamics

By the moment that we are going to use the sequential binary switch as a model for our memory, its dynamics, according to what has been discussed in Sect. 4.5, will be represented by the time evolution of the probability density $p(x, t)$ in a double-well potential.

Let us assume we want to preserve over time a memory where the bit "1" is stored. Thus, the initial state of the memory is S_1. In this configuration (Fig. 6.1b), $p(x, 0)$ presents a sharp peak centred in $x > 0$. If let the probability density evolve, $p(x, t)$ will first relax inside the right well within the relaxation time τ_w and then it will diffuse into the left well of the potential, by growing a pick also in that well (Fig. 6.1b–f).

The probability $p_0(t) = \int_{-\infty}^{0} p(x, t)dx$ increases with time, and eventually, it will reach its equilibrium condition, $p_0 = ½$, where the memory is definitely lost (Fig. 6.1f). As we have seen, to fight this phenomenon, we perform at time t_R a refresh operation that takes a time $t_p \ll t_R$. Let's suppose that after a time t_R the system is still in the "1" (right) well. This refresh operation reshapes the peak of the probability density function into the initial $p(x, 0)$, the memory is thus preserved and the cycle starts again. We have assumed that the refresh operation finds the system still in the right well. Clearly, as the time passes, the probability of finding the system in the wrong well increases and so increases the probability of error P_E. At time t_R, before the first refresh occurs, we have $P_E = p_0(t_R)$, thus the probability of being still in the right well is $1 - p_0(t_R)$. If we want our memory to last for a time $\bar{t}$, then we need to perform this operation a number of times.

$$N = int\left(\bar{t}/t_R\right)$$

With *int (a)* that indicates the integer part of *a*. Being $\bar{t} \gg t_R$, *N* is usually a large number. After *N refresh operations*, the probability to be still in the right well is

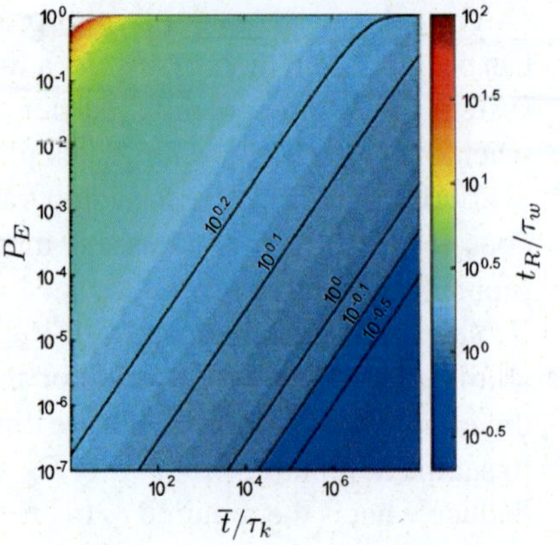

Fig. 6.2 t_R as function of $\bar{t}$ and P_E for a memory represented by an overdamped system in a bistable Duffing potentials. Here $\bar{t}$ is given in unit of the global thermalization time τ_k, while t_R is given as a multiple of τ_w, i.e. relaxation time within one well of the potential. The results here presented are obtained with $\tau_k = 5.3 \bullet 10^4\ \tau_w$

$(1 - p_O(t_R))^N$ and thus the probability of error will be:

$$P_E(\bar{t}, t_R) = 1 - (1 - p_O(t_R))^{\frac{\bar{t}}{t_R}} \tag{6.1}$$

From this relation, we can obtain $t_R = t_R(\bar{t}, P_E)$ that provides the maximum refreshing interval that satisfies the given conditions. In fact, if we wait to perform a *refresh operation* for a time interval longer than t_R, then the condition of keeping the probability of error below P_E is not met anymore.

The quantity t_R, that is of clear practical interest, can be obtained by solving the relevant Fokker–Planck equation for $p(x, t)$ and, from this, by computing P_E according to (6.1) for different t_R values. In Fig. 6.2, we have computed [1] t_R from an initial condition, where $p(x, 0)$ is a Gaussian function with standard deviation σ_i:

$$p(x, 0) = \frac{1}{\sigma_i \sqrt{2\pi}} e^{-\frac{(x-1)^2}{2\sigma_i^2}}$$

As it is well apparent, the larger we select $\bar{t}$ and the smaller P_E, the shorter is t_R, meaning that we need to perform the *refresh operation* more often.

6.3 Fundamental Energy Limits for Memory Storage

We are now in position to inquire what are the fundamental limits, if any, in the energy consumption for memory storage.

Let's start with memory writing. According to our model, as we have seen in Sect. 5.4, if our sequential switch is at equilibrium, in order to write one bit, we need to perform a Landauer reset towards S_0 or S_1. This operation costs at least the Landauer fundamental limit, with all the caveats that we discussed there. On the other hand, if our system is not relaxed at global equilibrium, then the writing of a single bit can be accomplished by performing a switch operation from S_0 to S_1 or S_1 to S_0, depending on the bit that is needed and the initial condition of our system. This operation, for what we have seen in Chap. 5, can be performed without spending any energy.

Once the memory has been written, there is the periodic *refresh operat*ion to be performed. What is the minimum energy required to accomplish the refresh? Based on what we said before, the refresh operation consists in transforming $p(x, t_R)$ into $p(x, 0)$ inside the single well that encodes the bit. We also assume that $t_R \ll \tau_K$ holds, so we can approximate the dynamics in each well with that of a harmonic oscillator near a potential minima x_m. With these approximations, we can locally approximate the real $p(x, t)$ with a Gaussian peak centred around x_m with a standard deviation $\sigma(t)$ that changes as the refresh takes place [2].

From this perspective, the refresh operation consists in applying an external force whose effect is to change $\sigma_f = \sigma(t_R)$ into $\sigma_i = \sigma(0)$. The latter is a design parameter, while the former depends on the intrawell relaxation mechanisms and, in the harmonic approximation, it is given by [1]

$$\sigma(t_R) = \sqrt{\sigma_w^2 + e^{-\frac{t_R}{\tau_W}}(\sigma_i^2 - \sigma_w^2)}$$

where σ_w indicates the equilibrium standard deviation in the harmonic potential.

What is the minimum energy required to operate this refreshing procedure? The transformation of the probability density, within our approximation, is the simple reduction of the standard deviation from $\sigma_f = \sigma(t_R)$ to $\sigma_i = \sigma(0)$. This clearly implies a squeezing of the $p(x, t)$ (See Fig. 6.3) that causes a reduction of the system entropy for a quantity $\Delta S = S(t = 0) - S(t = t_R) < 0$.

By the moment that the refresh operation requires a decrease in the system entropy, the energy has to be provided in terms of work done on the system $Q = T\,\Delta S$. This minimum amount of energy required by the entropy variation is clearly a function of the spread of the probability density $p(x, t)$ occurring during the time t_R. If we model the system entropy according to Shanon with.

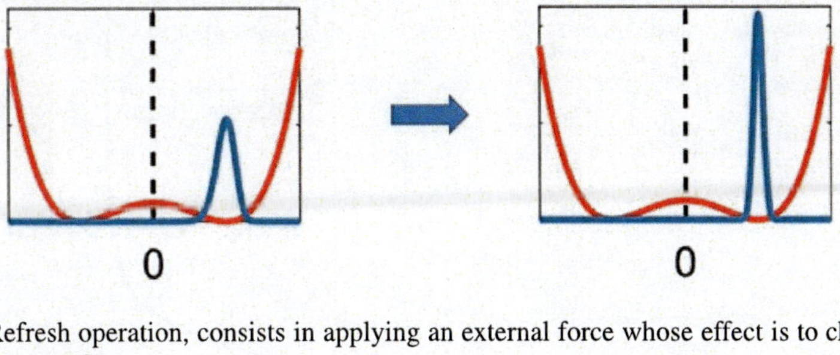

Fig. 6.3 Refresh operation, consists in applying an external force whose effect is to change $\sigma_f = \sigma(t_R)$ into $\sigma_i = \sigma(0)$

$$S(t) = -\int_{-\infty}^{+\infty} p(x,t)ln(p(x,t))dx,$$

we can express ΔS in terms of the ratio $\frac{\sigma_f}{\sigma_I}$ as [1]:

$$\Delta S = k_B ln\left(\frac{\sigma_f}{\sigma_i}\right)$$

Thus, putting together the various quantities that we have found, we have

$$Q(t_R) = -NT\Delta S = \frac{\bar{t}}{t_R} k_B T ln\left(\frac{\sqrt{\sigma_w^2 + e^{-\frac{t_R}{\tau_w}}(\sigma_i^2 - \sigma_w^2)}}{\sigma_i}\right)$$

This is the expression of the minimum energy to be dissipated during the refresh procedure, if the procedure is performed N subsequent times at t_R intervals.

In Fig. 6.4, we show the minimum energy $Q(t_R)$ for a given choice of P_E and $\bar{t}$. It is interesting to observe that this is an increasing function of t_R that shows two relevant features:

(i) $Q(t_R)$ approaches the value 0 when t_R goes to 0.
(ii) $Q(t_R)$ grows to infinity when t_R approaches a finite value t_{RMax}.

The feature (i) indicates that it is indeed possible to preserve the memory for a time $\bar{t}$ with probability P_E by spending zero energy. Clearly, this is possible as t_R approaches 0. This implies that between two subsequent refreshes, the entropy has no time to grow because the probability density has no time to change significantly. Such a condition, although not forbidden by the law of physics, raises some practical issues as it implies a continuous refreshing process with no memory availability for use.

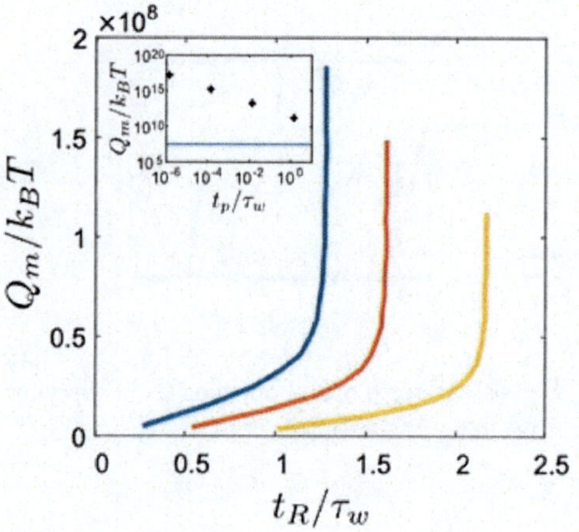

Fig. 6.4 Plots of the heat Q required to preserve the memory for $\bar{t} = 10^3 \tau_k$ as a function of t_R. Blue lines are obtained with $P_E = 10^{-6}$, red lines with $P_E = 10^{-4}$, yellow lines with $P_E = 10^{-2}$ [1]

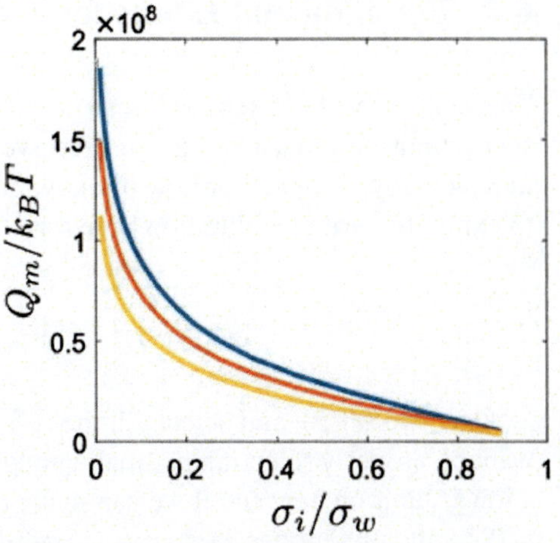

Fig. 6.5 Plots of the heat Q required to preserve the memory for $\bar{t} = 10^3 \tau_k$ as a function of σ_i. Blue lines are obtained with $P_E = 10^{-6}$, red lines with $P_E = 10^{-4}$, yellow lines with $P_E = 10^{-2}$ [1]

The divergence in (ii) can be understood as a consequence of (6.1).

As a matter of fact when we set our desired P_E, that is the probability of memory loss after the time $\bar{t} > t_R$, this cannot be smaller than p_0 that is the probability of memory loss at t_R. As a consequence, it does exist a maximum refresh time t_{RMax} above which we cannot satisfy the required condition anymore.

On the other hand, we can look at the behaviour of the minimum energy Q by another perspective. In Fig. 6.5, we show the minimum energy $Q(\sigma_i)$ for a given choice of P_E and $\bar{t}$. This is a decreasing function of σ_i that shows two relevant features:

(iii) It reaches the value 0 when σ_i goes to σ_w
(iv) It grows to infinity when σ_i approaches 0.

Condition (iii) indicates, as stated above, that it is possible to preserve the memory for a time $\bar{t}$ with probability P_E by spending zero energy, and this is realized when σ_i approaches σ_w. This is because when $\sigma_i \cong \sigma_w$ the system dynamics is at equilibrium and the entropy does not grow anymore (for some time). It will start growing again as soon as a new peak is formed on the left well.

On the other end, condition (iv) implies that the amount of energy required to keep the memory, grows boundless when we select an arbitrarily small confinement for the initial probability density ($\sigma_i \simeq 0$). This last condition, in any real physical system, is limited by the Heisenberg indetermination principle that sets a lower bound [3] for the initial standard deviation σ_i, i.e. $\sigma_i \geq \sigma_{iMin}$. As a consequence, the amount of energy required reaches a finite maximum.

In conclusion, we have seen that that in order to preserve a memory for a given time with a given error probability, it is possible indeed to spend an arbitrarily little amount of energy, provided that the refresh procedure is performed arbitrarily often (condition i) or arbitrarily close to thermal equilibrium (condition iii).

6.4 Fundamental Limits for Memory Preservation

The existence of existence of a finite σ_{iMin} has another important consequence: it sets a limit on our capability to preserve a given memory forever, no matter how much energy we spend or how often we refresh. This is apparent when we use (6.1) to explicitly write the duration time $\bar{t}$ as

$$\bar{t} = t_R \frac{\ln(1 - P_E)}{\ln(1 - p_0)}$$

Once we set P_E and select a finite refresh time t_R, we can make $\bar{t}$ as large as we want by properly selecting p_0 small enough. However, the existence of $\sigma_{iMin} > 0$ sets a lower limit on how small we can make p_0.

Specifically, it does exist a p_{0Min} and this is when the probability density p(x, t) is squeezed at its maximum in the bit "1" well. Why can't we make p(x, t) infinitely squeezed, like a delta function? The answer is in the Heisenberg indetermination principle that requires for the position x(t) to obey the uncertainty relation

$$\sigma_x \sigma_p \geq \frac{h}{4\pi}$$

where σ_x is standard deviation of the position and σ_p is the standard deviation of the momentum $p = mv$. If we operate our memory at thermal equilibrium, as it is customary, then we can express σ_p in terms of the equipartition theorem:

$$\sigma_p = m\sqrt{\langle v^2 \rangle - \langle v \rangle^2} = \sqrt{mk_B T}$$

Thus

$$\sigma_x \geq \frac{h}{4\pi} \frac{1}{\sigma_p}$$

and

$$\sigma_{iMin} = \frac{h}{4\pi} \frac{1}{\sqrt{mk_B T}}$$

The existence of a $\sigma_{iMin} > 0$ implies that even at t = 0, it is not possible to have $p_0 = 0$. This shows that the condition $p_0 = 0$ and $p_1 = 1$, when we write a memory cannot be exactly satisfied, as we implied in 6.1. If we indicate with p_{0Min} the initial minimum probability, then the maximum memory duration will be expressed as:

$$\bar{t}_{Max} = t_R \frac{ln(1 - P_E)}{ln(1 - p_{0Min})}$$

In practical memories, this time limit $\bar{t}_{\text{Max}}$ can be very large. As an example, we consider a micromechanical memory device [4]. If we assume the distance between the two wells $x_m = 1\ 10^{-9}$ m and a refresh period $t_R = 6.6\ 10^{-3}$ s, we have $\sigma_{iMin} = 9.6\ 10^{-20}$ m. If we set $P_E = 1\ 10^{-6}$ we have $\bar{t}_{\text{Max}} = 5.9\ 10^7$ s (approximately 2 years). On the other hand, if we set $P_E = 1\ 10^{-4}$ we have $\bar{t}_{\text{Max}} = 5.9\ 10^9$ s (approximately 200 years).

In conclusion, we have shown that the Heisenberg indetermination principle, that prescribes a maximum possible confinement of the relevant observable, implies the existence of a finite minimum uncertainty during the writing operation. This uncertainty reflects in the existence of a minimum width for the initial probability density of the memory device. The most dramatic consequence of this minimum width is that refresh strategies will inevitably fail after a maximum time given by the fundamental laws of physics. In other terms: no matter how much energy we invest, how often we refresh or how close to equilibrium we operate, there is a finite time after which any memory will be irremediably lost.

6.5 Beyond Digital Memories

Most of the reasoning developed for digital memories remains generally valid also for others kind of memories, like human memory, for example.

The preservation of our biological memory is of fundamental importance in human life, and its loss is a cause of great pain, as in the case of dementia for the elderly. However, even in the case of a perfectly healthy person, no matter how important the object of remembrance is, it is a common experience that memories tend to weaken over time. Probably the first who systematically studied this phenomenon was Hermann Ebbinghaus (1850–1909), a Prussian philosopher who conducted a series of studies on the persistence of human memory, eventually published in a 1885 book entitled "Memory: A Contribution to Experimental Psychology". Ebbinghaus discovered the "forgetting curve", an exponentially decreasing curve that shows that the ability to remember, unless continuously reinforced, tends to halve every hour or so. The reinforcement process studied by Ebbinghaus closely resembles the *refresh operation* performed on digital memories.

Not only human memory is doomed to dissolve with time, but also material objects, artefacts, monuments and landscapes require periodic maintenance to counterbalance the effect of time which modifies shapes and levels differences, following the universal tendency of entropy to increase.

The existence of a maximum duration time for any possible storage medium might come as a surprise. On the other hand, we are used to ancient artefacts and paintings that make up the memory of bygone times. The law of the finite duration of memory, of every memory, also applies to them. They are analogue memories that record information encoded by other men in other times.

Each form of memory has its necessary refresh period and its maximum duration time, depending on the specific nature of the material support in which it is made. It can range between few seconds, for electronic digital memories if no refresh is performed, to (probably) million years for carefully preserved monuments.

Will we be able to preserve the large amount of digital data our society is producing? And if so for how long? We must be conscious that if we decide to build a future society on digital data, then it is best to be aware that it will not last forever: the more volatile and error-prone our memories are, the shorter their duration will be.

References

1. D. Chiuchiù, M. López-Suárez, I. Neri, M. C. Diamantini, L. Gammaitoni, Cost of remembering a bit of information. Phys. Rev. A **97**, 052108 (2018)
2. D. Chiuchiú, Time-dependent study of bit reset. EPL **109**, 30002 (2015)
3. M.C. Diamantini, L. Gammaitoni, C.A. Trugenberger, Landauer bound for analog computing systems. Phys. Rev. E **94**(2016). https://doi.org/10.1103/PhysRevE.94.012139
4. I. Neri, M. López-Suárez, Heat production and error probability relation in Landauer reset at effective temperature. Sci. Rep. **6**(34039) (2016). https://doi.org/10.1038/srep34039

Chapter 7
What About Information?

Contra factum non valet argumentum.
Old Latin motto

Abstract In the previous chapters, we developed the basic foundational ingredients of the physics of computing by dealing with binary switches, their functioning and the fundamental limits that physics imposes. We have done all this without any reference to one of the most popular concepts that dominated computer science for almost a century: the notion of *information*. How is this possible? Don't we need *information* to describe the functioning of computing devices? In this chapter, finally, we discuss these aspects.

7.1 The Big Picture

In the previous chapters, we have slowly and carefully climbed the mountain of our "physics of computing" knowledge. Now we can collect our reward: we can finally see further than when we started our climb.

All automatic digital computing activities are made with elementary devices that we call combinational binary switches. These devices can be combined together to make logic gates and memories, that are the basic components of modern computers, according to the von Neumann architecture. We have shown that all the computing activity with these devices can be performed without spending any energy.

We have shown that digital memories can be made with sequential binary switches. The operation of these devices can be performed without spending any energy, with the exception of the reset operation that can be necessary once the device has relaxed at thermal equilibrium. We have shown that maintaining the status of a digital memory can be performed by spending and arbitrarily small amount of energy, if the refresh operation is performed often enough or arbitrarily close to its local equilibrium. We have also shown that, no matter what we do, there is no way to keep the memory status forever, due to unavoidable growth of the memorization error over time.

L. Gammaitoni, *The Physics of Computing*,
https://doi.org/10.1007/978-3-030-87108-6_7

These are the fundamental limits in the physics of computing, regarding energy consumption, and they have been established on the base of the laws of physics as developed within statistical classical and quantum mechanics, with specific reference to the set of tools developed within the so-called stochastic thermodynamics approach.

The very topic of "energy limits in computation" has been the subject of a long debate in the vast scientific community [1]. The controversy has developed mainly after the connection between energy and the notion of *information.* Before dealing with these aspects, we will go through a brief introduction on the notion of *information.*

7.2 Information: What is It?

The notion of *information*, the way we use it today, was introduced in the scientific literature by Claude Shannon in a paper published in 1948 and entitled "A Mathematical Theory of Communication" [2]. In a time in which telephone system was booming, Shannon's aim was to advance a general theory of signal transmission, capitalizing on the work done earlier by Harry Nyquist and Ralph Hartley (1888–1970). *Information* is a quantitative label that can be associated to a message to be transmitted. More specifically, it is a number that represents the degree of uncertainty of receiving a message, based on the known probability distribution of the possible messages.

A popular example of computation of such a number is the following: let's suppose that we are waiting for an answer to a question that we have advanced, something like "will you marry me?". Now, if we know (possibly by experience!) that we get a negative answer 90% of the time and positive 10% of the time, then the message that we are getting will carry certain amount of *information.* Let's suppose now that the odds are different: positive answer and negative answer same probability, 50% each. In this second case, the uncertainty on the outcome of the answer is maximum and thus it will also be maximum the quantity of *information* associated with it. If we call H this quantity, and p_i the probability of receiving message number i, out of M existing messages, then

$$H = -\sum_{i=1}^{M} p_i \, log_a(p_i)$$

a is a generic base and H is clearly a positive number, being $p < 1$.

We can also use H to compare different messages. For example, getting the news of the outcome of a coin flip (head and tail with same probability, H_c) is different from getting the news of the outcome of a roll of a die (with six equally likely outcomes, H_d). We have $H_d > H_c$.

If instead of sending messages we are sending symbols, like the alphabet characters, the same rule applies. In a communication composed by M possible symbols, each of them with a given p_i probability and appearing in the communication n_i times, the quantity of information will be:

$$H = -\sum_{i=1}^{M} n_i p_i \, log_a(p_i)$$

On the other hand, if we are using bits (binary digits) to send the message, we will have only two different symbols "0" and "1". In a long message with N bits, it is very common to get a number of "0" and "1" that is approximately equal, i.e. we have the same probability to get any of them: $p_0 = p_1 = ½$. Thus, the quantity of *information* per bit is

$$H = -\sum_{i=1}^{2} \frac{1}{2} \, log_a\left(\frac{1}{2}\right) = 1$$

where we have chosen $a = 2$. In such messages, each bit carries H = 1 of information. The choice of the base of the logarithmic function sets the units for the *quantity of information.* For $a = 2$ we have *bits*, $a = e$ we have *nats* and $a = 10$ we have *dits.* In a message composed by N bits, we will have N bits of *information.*

The concepts developed by Shannon in the 40s of last century were aimed at establishing a mathematical theory for finding the best way to transmit a message from one point to another, in the unavoidable presence of the noise. Questions like: how much information can we send through a channel with a given signal-to-noise ratio? What is the best code for transforming a message into a more robust form for transmission? Eventually, Shannon worked also on the best way to encrypt a message in order to make it resistant to eavesdropping. For what we are concerned in this book we will not proceed further in the theory of information, but in the next paragraph, we will look into a specific application of this theory to our problem of establishing the energy limits in computation.

7.3 Landauer's Principle: Long Story Short

We have already mentioned the work published by Ralph Landauer in 1961 [3]. In this work, in addition to the discussion of the bit reset, Landauer introduced the notion of logical irreversibility, as applied to digital computing devices, by using the notion of *information.*

Landauer writes:

> We shall call a device logically irreversible if the output of a device does not uniquely define the inputs.

Fig. 7.1 A computation device is ideally represented as an information input–output device

According to this definition, a computation device is ideally represented as an input–output device (like the adder in Fig. 2.11) with some logic values entering the device and some logic values leaving the device. The input and the output terminals being able to assume logic values can be characterized for the amount of information that they carry (see Fig. 7.1). Typically, they carry one bit per terminal. As an example, in an ordinary OR gate (see Fig. 2.10) we have two input terminals and one output terminal, thus we have two bits in and one bit out. In a situation like this, a necessary condition for assuring the logical reversibility is to have a number of inputs not larger than the number of outputs. An OR gate is a typical irreversible logic gate, as are AND, NAND and NOR. On the other hand, the NOT gate (one input and one output) is a logically reversible gate, because from the logic state of the output terminal is always possible to reconstruct the logic state of the input terminal.

In the same passage, Landauer writes:

> We believe that devices exhibiting logical irreversibility are essential to computing.

In a sense they are, as we have discussed in Chap. 2, as far as we keep using traditional logic gates. However, as Landauer himself and, soon after, Charles Bennett (at that time at IBM) have shown, all the computation can also be made by perfectly reversible logic gates [4]. Examples of these gates have been proposed by Edward Fredkin and Tommaso Toffoli in the early 80s of last century.

Finally, Landauer adds:

> Logical irreversibility, we believe, in turn implies physical irreversibility, and the latter is accompanied by dissipative effects.

Undoubtedly, this is the most far-reaching result that Landauer is claiming. According to this statement, a logically irreversible device cannot be physically operated without spending energy, because dissipative effects are unavoidable. Landauer's belief originates from the idea that the decrease of information between input and output, an operation that is sometime addressed as "*information* erasure", is necessarily connected to a decrease of physical entropy in the physical system that embodies such a device.

Such a connection between logical irreversibility and physical irreversibility has been at the centre of a very long and controversial debate, not free from misunderstandings and oversights. It is our opinion that, in this respect, Landauer was wrong: logical irreversibility does not imply physical irreversibility, and to make this point clear, we made an experiment, whose results have been published in 2016 [5]. In the next paragraph, we will present the main aspects of the experiment and show how it denies Landauer's claim.

Here, we observe that the conceptual approach adopted by Landauer, although very popular in the *information theory* framework, has no direct connection

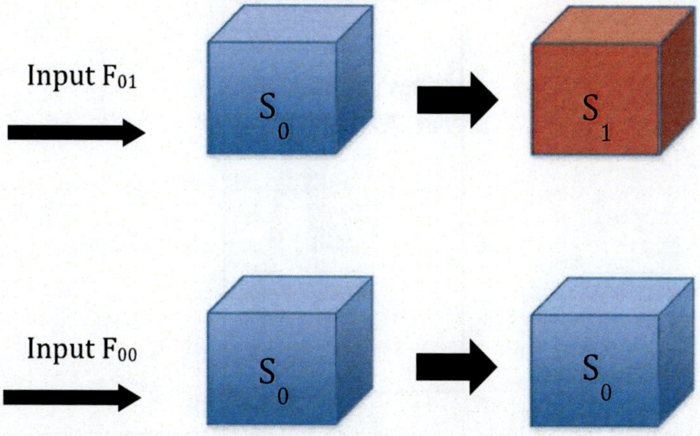

Fig. 7.2 Here the input is represented by the resultant of the applied forces, and the output is the physical state of the system: S_0 or S_1

with a typical physical model of the computing device, made according to the thermodynamics.

Following what we have seen in Chap. 2, a computation is the result of a **change of state** of a given physical system, under the action of an external force (see Fig. 2.4, here reproduced as Fig. 7.2). According to Landauer, the dof bearing *information* is represented by the status S_i of the system and by the input forces, while in a thermodynamic picture the change in entropy (that is a state variable) is uniquely related to the system itself and the forces are not accounting for any entropic contribution.

Although Shannon *information* and Gibbs entropy have the same formal aspect, they are not the same. The first is a mathematical quantity associated with the probability of a message, while the second is related to the probability of a microstate in a physical system. While the first has no connection with the physical nature of the message, the second is a direct consequence of the nature of the physical system itself. Thus, it seems natural to conclude that while Gibbs entropy is associated with the way the energy is exchanged according to the second principle of thermodynamics, Shannon information is not.

7.4 Sub-k_BT Microelectromechanical Irreversible OR Logic Gate

In this paragraph, we present a brief account of an experiment made by Miquel Lopez-Suarez, Igor Neri and Luca Gammaitoni, at the NiPS Laboratory, in Perugia, in 2015 and published on Nature Communication in 2016 (see reference).

The goal of the experiment is to show that an OR logic gate, a well-known logically irreversible device, realized with a microelectromechanical cantilever, can be

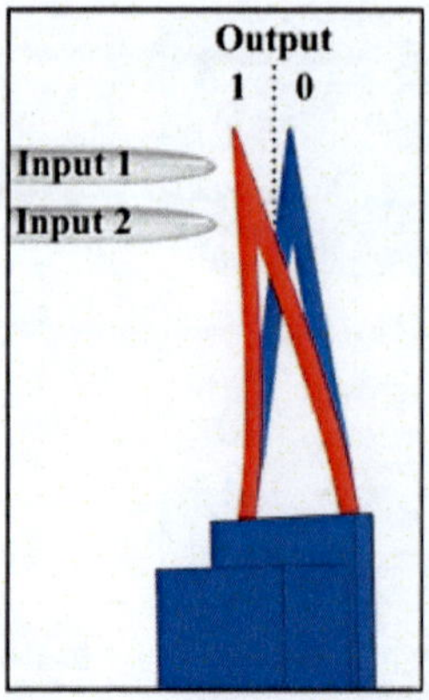

Fig. 7.3 Scheme of the OR logic gate made with a mechanical cantilever. Inputs are the forces acting on the cantilever through the electrodes. The position of the cantilever tip encodes the output of the logic gate

operated with energy well below the expected Landauer's limit $k_B T\, ln(2)$, provided the operation is slow enough and frictional phenomena are properly addressed.

The experimental apparatus employed is similar to what we have discussed in Sect. 5.2, and we will reproduce here only the most relevant aspects that are different.

In this case, we have a single combinational switch made with a Si_3N_4 cantilever that can be bent by applying electrostatic forces with **two** electrical probes close to the cantilever tip (see Fig. 7.3). The logic output of this device is encoded in the tip position as shown in the figure. The logic values of the two inputs are associated with the voltage of the respective probe. In the following, with the letter *V* we will indicate the voltage and not the potential energy as in the other paragraphs.

The principle of operation of this device is similar to what we would have in an OR logic gate realized with transistors (see Fig. 7.4). When the bias voltage input is applied to the transistor gate, the channel is opened and the conducting state of the transistor changes the voltage of the output terminal. In our cantilever, once the electrostatic force is applied, the position of the cantilever is changed (bent position). In the transistor, once the bias voltage input is zeroed, the channel closes and the initial state is recovered. The same happens in our cantilever: once the force is removed, the cantilever position goes back to the zero-force state (vertical position).

One relevant difference between the cantilever-based device, considered in this experiment, and transistor-based logic gates is that, for the former, input and output have different natures (electrostatic forces and mechanical displacement respectively), while for the latter, both are voltages. This is clearly relevant for combining together multiple logic gates; however, it has no relevancy for the purposes of this experiment.

We can operate our cantilever-based device as an OR gate, by carefully setting the threshold for the two states S_0 and S_1. We illustrate in Fig. 7.5 the statistical distribution of the position of the cantilever tip, as a function of the inputs. We observe

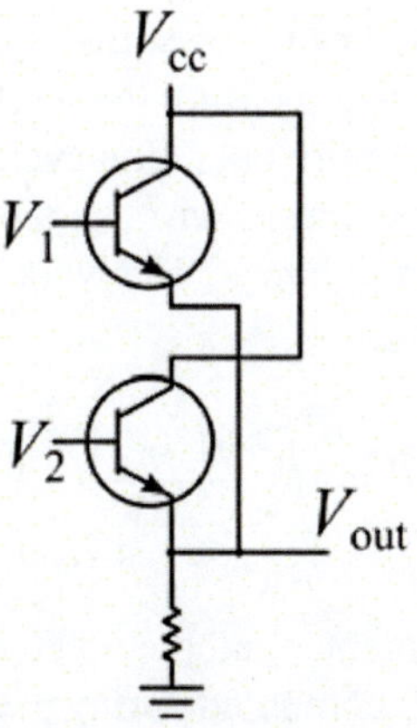

Fig. 7.4 Scheme of the OR logic gate made with two cascaded transistors. Inputs are the voltages entering the transistor base. The output is represented by the voltage at V_{out}

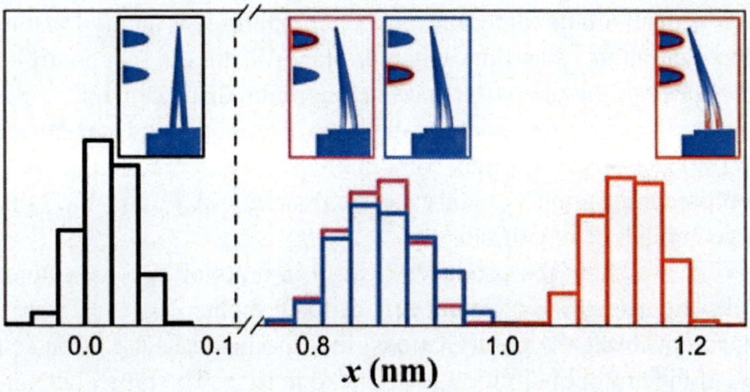

Fig. 7.5 Statistical distribution of the position of the cantilever tip, as a function of the inputs. A red contour on the probe indicates that electrostatic force is applied. Tip displacement is measured in nm

that logic states "0" associated to S_0 and "1" associated to S_1 are clearly distinguished, since the overlap between the corresponding probability density functions is negligible. Moreover, logic inputs corresponding to the mixed states (I1 = 1, I2 = 0 and I1 = 0, I2 = 1) produce a physically indistinguishable position distribution and thus the same state S_1. Finally, the logic input I1 = 1, I2 = 1 produces a larger displacement that, due to the threshold setting, belongs to the same logic partition of S_1. Under these circumstances, the cantilever-based gate performs like an OR gate that is a logically irreversible device: in fact, there is at least one case where, from the sole knowledge of the logic (and the physical) output, it is not possible to infer the status of the logical inputs.

The goal of the experiment is to measure the energy dissipated during the OR gate operation and to show that this energy can be made arbitrarily small and, certainly smaller than *ln*(2) $k_B T$, as expected according to the Landauer's principle.

The experiment is performed as follows: the logic gate is switched from state S_0 to S_1 and from state S_1 to S_0. We measure the work L performed on the system by the external forces during this entire operation cycle. Since, in a cycle the change in the internal energy U of the system is zero, we have $Q = L$, i.e. all the work done on the system is dissipated through heat. The work done is obtained, as in Sect. 5.2, through a Stratonovich integral

$$L = \int_0^{t_p} \frac{\partial U(x, V)}{\partial V} \dot{V} dt$$

by measuring the tip displacement x and the voltage V applied to the probes.

The Stratonovich integral is computed during a single operation cycle of duration τ_p. During this cycle (Fig. 7.6) if the ith logical input I_i $(i = 1, 2)$ is set from 0 to 1, the voltage of the ith probe increases linearly from $V = 0$ to $V = V_0$ during a time interval of length $\tau_p/8$. The same time interval occurs when decreasing the voltage from $V = V_0$ to $V = 0$ to set the input from the logic 1 to 0. In this experiment, every operation cycle starts with all inputs set to 0; hence, no initial force is applied to the cantilever. The protocol duration τ_p and maximum voltage V_0 are the two control parameters that set the speed of the operation and the maximum displacement, respectively.

To measure the energy dissipation, the logic gate is operated at room temperature, addressing all the possible input combinations. Each combination is applied for different protocol duration τ_p. For each combination of τ_p and V_0, Q is evaluated over a large number of repetitions.

In Fig. 7.7, we show the produced heat Q in units of $k_B T$ as a function of the protocol duration for three different sets of logic inputs: '01', '10' and '11'. As it is well apparent in all the combinations, the dissipated heat can be reduced well below $k_B T$ if the protocol duration is extended in time. The typical exponential-like decrease of the dissipated heat can be explained in terms of a proper dissipation model due to internal friction happening in the cantilever material. Detail on the model is discussed in the cited article.

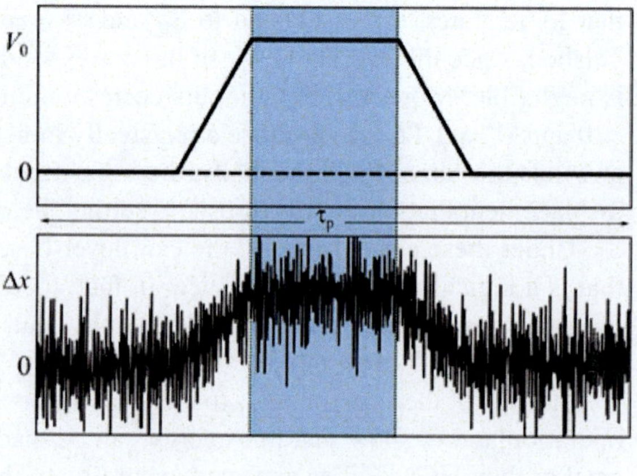

Fig. 7.6 Value of the electrode voltage V (logic input) and the consequent displacement of the cantilever tip x (gate output)

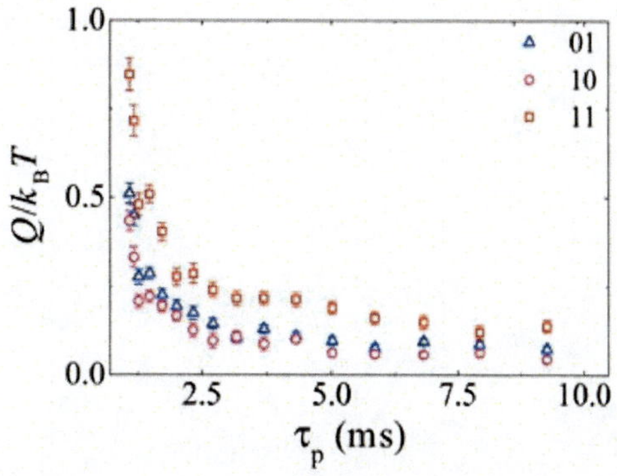

Fig. 7.7 Average produced heat as function of protocol time τ_p, for the three different sets of logic inputs: "01", "10" and "11". With increasing the duration of the protocol time, the produced heat decreases following a power law

In conclusion, we believe that this experiment rules out the presence of a finite "minimum dissipated heat" due to logical irreversibility, an argument often invoked when the reduction of input–output information is considered. We stress here that our experiment does not question the so-called Landauer reset operation, where a net decrease of physical entropy requires a minimum energy expenditure, as we have seen in the previous chapter. What we have here is a logically irreversible computation, that is a generic process where a decrease in the amount of information between the output and the input is realized with an arbitrarily small energy dissipation; this shows that logical reversibility and physical reversibility have to be treated on independent bases.

References

1. C.S. Lent et al., *Energy Limits in Computation.* (Springer, 2019)
2. C. Shannon, A mathematical theory of communication. Bell Syst. Tech. J. **27**, 3 (1948)
3. R. Landauer, Dissipation and heat generation in the computing process. IBM J. Res. Dev. **5**, 183–191 (1961)
4. C.H. Bennett, Logical reversibility of computation. IBM J. Res. Dev. **17**(6), 525–532 (1973)
5. M. López-Suárez, I. Neri, L. Gammaitoni, Sub-kBT micro-electromechanical irreversible logic gate. Nat. Commun. **7**, 12068 (2016)

Chapter 8
Cloud Computing and Data Transmission

There are no rules of architecture for a castle in the clouds.
Gilbert K. Chesterton

Abstract With the fast development of Internet, in the last thirty years we have assisted at an increasing number of computing services that have been progressively moved from the computing device itself to other, usually most powerful computers, remotely located in some highly efficient computer structure. The data exchange between the computing device and such remote computers is managed by high-speed communications through dedicated communication lines. In this chapter, we discuss some of the physics aspects associate with such a computing modality, where communication plays a significant role.

8.1 A New Name for an Old Situation

Cloud computing is a new name for a not-so-new concept. In its simplest terms, *cloud computing* means no more that doing computation by utilizing/sharing some resources, like memory, computational power or other services, located elsewhere and accessible via the Internet (or any specific network connection). As such, the idea of *cloud computing* has been around since the early stage of Internet, in the 70s of last century, and is somehow intrinsically connected with the idea of the network itself.

A typical example of such a computing organization is the service called Siri (by Apple) or Alexa (by Amazon). Every time that we pronounce the sentence "Hey Siri" and then ask something, we are connected with the Apple servers that analyse our question. How does this work? Simple [1]: the microphone in our phone is always on and samples the sounds that it receives at a rate of 16 kHz. The sampled sounds are organized into "sound frames" 0.01 s long (approximately 160 samples). Our device collects 20 of these frames and analyse them with a "simple" classification algorithm (deep neural network) with the goal of discovering if these frames might refer to the

L. Gammaitoni, *The Physics of Computing*,
https://doi.org/10.1007/978-3-030-87108-6_8

sound "hey Siri" as pronounced by a human voice. This is what is called the *voice trigger* and runs locally, on the device itself. If the sentence "hey Siri" is detected, the following samples are sent, through the Internet, to the Apple servers that keep analysing the samples in order to detect the request and send back an appropriate response. If not, the samples are ignored.

Analogously, Alexa is Amazon's cloud-based voice service available on many devices from Amazon or third parties that use voice recognition algorithms through the Internet.

These are, clearly, just two of many applications, and they strictly relay on the existence of the communications through Internet to function. The reason is that the computational heavy tasks required by these services cannot be performed directly on-board of the device. The hidden condition for such services to function is the existence of a quasi-free, fast and ubiquitous data communication networks. The physics involved here is the physics associated with signal transmission.

In the next paragraphs, we will discuss the physical limits in data communication. Once again, this is a topic where Claude Shannon was among the first to provide useful insights.

8.2 Signal Transmission

In its early paper "A mathematical theory of communication" that we cited in Chap. 7, Shannon introduces a model of communication that can be easily summarized in Fig. 8.1. In his model, all the communication process can be reduced to a point-to-point transmission of a signal that has been properly prepared after a message has been codified. The signal, during the transmission phase, is possibly corrupted by the addition of noise. Once received, the signal is properly decoded and made intelligible at the destination.

The point-to-point transmission of the signal employs a *communication channel.* This is a physical medium, like a metallic wire, a coaxial cable, an optic fibre, air or the vacuum itself, that has some relevant quantitative characteristics. The most important characteristics is the signal bandwidth B that represents the difference between the highest and lowest transmissible frequency.

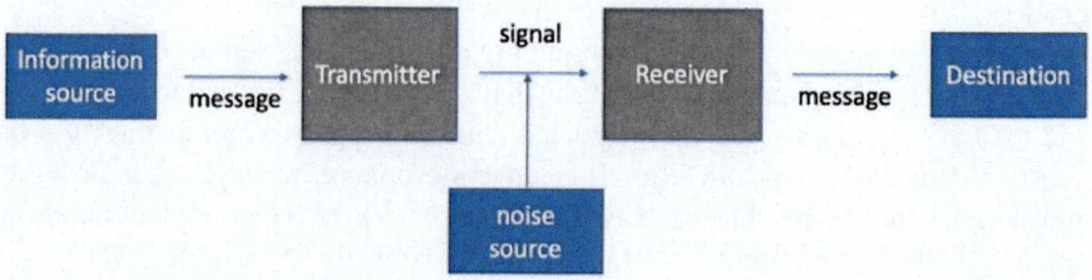

Fig. 8.1 Schematic representation of the communication process. Please note the role of noise that is explicitly acknowledged here

Fig. 8.2 Schematic representation of a simple transmission line

To better understand this concept, let's suppose to have an electric circuit like the one in Fig. 8.2. If we apply an alternate voltage generator V at one side of the circuit as input, we will observe a voltage V_c at the ends of the capacitor C, the output. This is a simple example of a transmission line, characterized by a resistance R and a capacity C, driven by a voltage $V(t) = V_o \cos(2\pi\nu t)$. The voltage $V_c(t)$ can be easily found to be a function of the frequency ν of the input, whose amplitude V_{co} scales with the factor (see Fig. 8.3):

$$A(\nu) = \frac{1}{\sqrt{1 + (2\pi \nu RC)^2}}$$

A simple interpretation of this behaviour is that not all frequencies in the input signal are transmitted with the same amplitude, with lower frequencies being transmitted better than higher ones (low-pass filter effect). The frequency where the transmitted power halves from its maximum is called *cut-off frequency* ν_c. Being the power proportional to the square of the amplitude, we have:

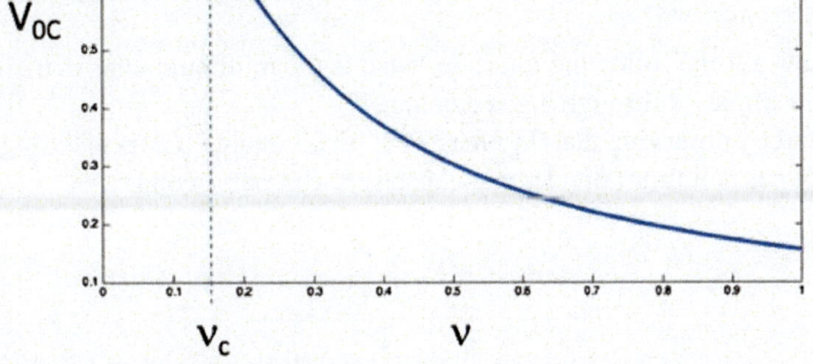

Fig. 8.3 Amplitude of the transmitted signal as a function of the input signal frequency. ν_c is the cut-off frequency that in this case marks also the bandwidth B

$$\frac{1}{\sqrt{1 + (2\pi \upsilon_c RC)^2}} = \frac{1}{\sqrt{2}}$$

By solving, we have:

$$\upsilon_c = \frac{1}{2\pi RC}$$

The transmission line bandwidth B is computed as the interval the transmitted power is above ½ of its maximum (see Fig. 8.3), thus $B = \nu_c - 0 = \nu_c$.

8.3 Minimum Energy for Transmitting One Bit

In this paragraph, we discuss the concept of information transmission and its implication on the amount of energy required. This topic has been addressed since the beginning of last century and has been put in the modern form by the father of information theory, Claude Shannon.

The starting point of Shannon's reasoning is the following: we want to transmit a certain message through a given communication channel and want that this message reaches its destination uncorrupted. The cause of potential corruption is called noise that might superimpose on the message. Shannon was able to demonstrate that if in a given channel characterized by a bandwidth B, at the destination you can measure an amount of noise power N and a signal power S, then the maximum number of bits per unit time (bps) you can transmit without corruption is:

$$C = B \log_2\left(1 + \frac{S}{N}\right)$$

C is called the *channel capacity,* and this relation is often addressed as the *Shannon–Hartley relation.* It is interesting to observe that C is directly proportional to the bandwidth B, that is a function of the physical nature of the signal and of the channel.

We now ask the following question: what is the minimum energy E_b required to transmit a single bit through a given channel?

We start by observing that the energy per bit is equal to the received signal power S (energy per unit time) divided per capacity C (bit per second):

$$E_b = \frac{S}{C}$$

On the other hand, the noise power N is equal to the noise spectral density N_0 (noise power per unitary bandwidth) times the bandwidth:

$$N_0 = \frac{N}{B}$$

Inserting these two into the *Shannon–Hartley relation*, we have

$$C_B = \frac{C}{B} = \log_2\left(1 + \frac{E_b}{N_0} C_B\right)$$

The quantity E_b is thus readily expressed as:

$$E_b = \frac{\left(2^{C_B} - 1\right)}{C_B} N_0$$

If the bandwidth B is much larger than the capacity C (meaning that we transfer bits very slowly), we can take the limit when C_B goes to zero. By the moment that

$$\lim_{x \to 0} \frac{(a^x - 1)}{x} = \ln a$$

we have

$$E_b = N_0 \ln 2$$

The energy per transmitted B is proportional to the noise N_0. This makes sense, because the larger is the noise, the larger has to be this energy in order to make the signal distinguishable from the noise. Thus, the minimum energy depends on how large is the noise. At the very minimum, even if we are able to get rid of all the other noises, the thermal noise at a given temperature T is an unavoidable source.

For the thermal noise, we have $N_0 = k_B\, T$. Thus, the minimum energy required for sending a bit of information, due to the fundamental thermal noise limit, is

$$E_b = k_B\, T \ln 2$$

As before, we would like to stress that this is fundamental limit that sets the minimum energy required. In practical systems, there are other noise sources that can play a relevant role as well, and the minimum energy is larger than this one.

One final word should be spent to compare this fundamental limit, i.e. the minimum energy required to transmit one bit, with the minimum energy required to do a switch event, i.e. the elementary step in the computation process. As we have seen, there is a minimum energy to be spent only in the case in which a Landauer reset is required. This amount of energy is dissipated in heat, and it is definitively lost during the computation. For the minimum energy required to send one bit, instead, E_b

is the energy associated with the physical signal that transmits the information. This quantity is not automatically dissipated in heat and, in principle, could be restored, once the signal is received at the destination.

Reference

1. https://machinelearning.apple.com/research/hey-siri

Chapter 9
Analog Computing

Check for updates

The real danger is not that computers will begin to think like men, but that men will begin to think like computers.
Sydney J. Harris

Abstract In Chap. 2, we discussed digital computing as a form of computing done with quantities represented by discrete units. In this chapter, we will discuss, instead, computing done with quantities represented by continuous values. How is this different? We show that computing with continuous quantities, usually called analog computing, is way more general than digital computing and can be performed with a vast variety of devices. However, when it comes to fundamental limits, still physics rules.

9.1 What is Analog Computing?

In digital computing, the quantities that we deal with are represented by rational numbers, expressed by a finite number of digits and thus embodied into physical quantities that vary stepwise assuming a finite number of distinct physical states, like binary switches. In analog computing, the quantities that we deal with are represented by physical quantities that vary continuously, like an electric current or voltage, to mention two of the most popular ones. The fact that an ordinary electric current, due to the corpuscular nature of the electrons, is not a truly continuous quantity will be ignored here, because its granularity can be neglected due to the smallness of the electric charge compared to the values of the currents normally used.

The world "analog" comes from the Latin "analogicus" and means *something that is similar to something else*. In this sense, *analog computing* is a computing that is accomplished by exploiting the analogy of the task at hand, with some physical phenomena.

There are at least two different types of analogies exploited: *direct analogy* and *indirect analogy*.

L. Gammaitoni, *The Physics of Computing*,
https://doi.org/10.1007/978-3-030-87108-6_9

In *direct analogy,* one exploits directly a physical property associated with the task to compute. As an example, if we want to compute the distribution of the loads necessary to obtain the stability of a building, we can make a reduced scale of the building itself and scaled weights, and adjust the weights in a try-and-error procedure, in order to find the proper loads.

In *indirect analogy*, one uses a physical property that is different from that involved in the task to be computed and thus it results in a more general and flexible approach. A typical example is a simple addition of numbers that can be accomplished with a slide ruler or an electronic circuit that can be built to solve a differential equation.

By the moment that analog computers based on *indirect analogy* are way more diffused and way more suitable for general purposes, in the following we will focus our discussion on them.

It is very important to understand how the principle of operation of an analog computer is very different from that of a digital computer. As we saw in Chap. 2, digital computing is based on the assembly of universal gates that perform all the arithmetic and logic operations by exchanging data with memories. Here, for a change, we have no memories whatsoever and all the data required during the computation have to be provided at the beginning of the process. In the next paragraph, we will see some examples taken from the quite general class of electronic analog computers and it will be clearer how the computing task is actually performed. For the moment, what we would like to stress is that in this case the typical standardization process, where all the computation is reduced to the use of the same task repeated again and again, that is, the base of digital computing, cannot be applied here, nor is the clocked procedure that rules digital computing. The abstract level of computation here is definitely brought down to the physics layer of the functioning of the device involved, and the fundamental physics principles behind analog computers are, more transparently than before, the very same fundamental physics principles that we are used to.

To provide a simple example of how an analog computer might work, let us look at Fig. 9.1. Here, we have a typical mass–spring model. The dynamic equation that rules the mass position can be expressed as:

$$m\ddot{x} + \gamma\dot{x} + kx = F(t)$$

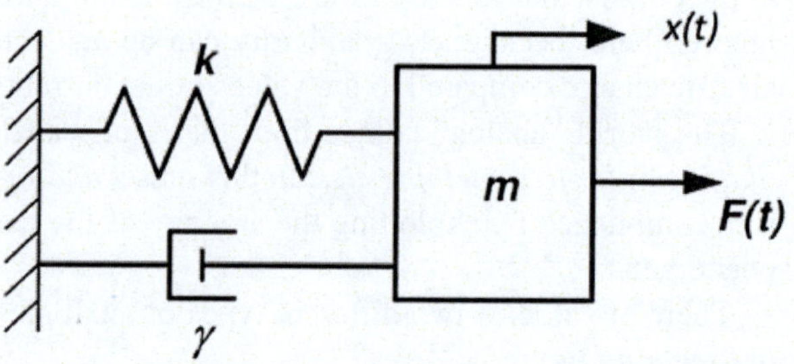

Fig. 9.1
Mass–spring–damper model

where there is no gravity involved, m is the inertial mass, γ is the (viscous) damping constant, k is the elastic constant and $F(t)$ is a time-dependent driving force.

By changing m into the inductance L, γ into the electrical resistance R, k into the capacitance C and $F(t)$ into the electromotive force $V_e(t)$, the same equation represents also the time evolution of the charge accumulated on the armatures of a capacitor. Moreover, the very same equation represents also the time evolution of the volume in an acoustic phenomenon, where m is replaced by the acoustic mass, γ by the acoustic resistance, k by the acoustic compliance and $F(t)$ by the pressure.

As we see, there are different phenomena that can be described by the same differential equation. If you look at this situation by a different point of view, you may realize that if you have a differential equation that you want to solve, all you need to do is to find the proper phenomenon that is described by that equation and then to perform an experiment of that phenomenon.

As an example, if you would like to solve the following equation

$$a\ddot{x} + b\dot{x} + cx = y(t),$$

you may build a mass–spring–damper experiment, with the proper values of mass, elastic constant, damping constant and driving force, and measure the value of the position x over time. Or, if you are more skilled with electric circuits, you can do the same in that ambit and so on in different areas. Your experimental apparatus is *de facto* an analog computer, built to solve a specific problem: how to integrate a given differential equation. Sometimes, this kind of computing is also called *analog simulation* because the experiment is built in order to simulate the expected time evolution of the differential equation solution.

9.2 Analog Computers Came First

As we have discussed in Sect. 1.3, devices employed for computation have been traced back to 40,000 BC. They were clearly analog computers, according to what we have seen in the previous paragraph.

The first mechanical device devoted to complex calculations, in modern sense, is probably the Antikythera mechanism (Fig. 9.2), an orrery designed to calculate astronomical positions. It was discovered in 1901 off the Greek island of Antikythera and has been dated to c. 100 BC. The device is surprisingly complex and looks much more modern compared to the technology of the time. It has 37 meshing bronze gears enabling it to follow the movements of the Moon and the Sun, to predict eclipses and to model the orbits.

To find a mechanical analog computer of a similar complexity, we have to jump to 1235 when Abi Bakr of Isfahan, Persia, invented an astrolabe incorporating a mechanical calendar and gearwheels. Another important analog computer, precursor of the already mentioned slide ruler, was the *proportional eight-pointed compass*,

Fig. 9.2 Antikythera mechanism [1]. It dates back to approximately 100 BC

a triangle with movable arms and cursor, invented in 1567 by Fabrizio Mordente (1532–1608), an Italian scientist. Galileo Galilei himself produced his own version of this military compass and trained students on its use. In the hall of fame of analog computer, a special place is devoted to Lord Kelvin (William Thomson) who invented in 1872 a mechanical machine employed in tide prediction. At the end of 1800s, a mechanical general-purpose analog computer was proposed by James Thomson, brother of Lord Kelvin, who proposed a machine specialized in solving differential equations (the *differential analyser*) based on wheel-and-disc mechanisms.

The diffusion of electricity and its use to power mechanical machines gave new impulse to the creation of electromechanical analog computers. In 1912, Arthur Pollen, an English businessman, developed a machine for computing the best trajectories for firing cannons aboard ships. Analog computer based on *direct analogy* was developed in the 1920s to analyse the newly born electric networks, by simulating them with small-scale circuital elements.

The Second World War was also the occasion for great impulse in the making of analog computers, mainly for military applications, especially in rocket launch and airplane flight simulation. Post-war efforts in science were multiplied, especially in the USA, and a number of initiatives took advantage of analog computers. Among these, there was the FERMIAC, a machine invented by Enrico Fermi (1901–1954) in

Chicago in 1947, to study the neutron dynamics. In 1950, Reeves Instrument Corporation, a weapon maker company, developed for the US military the *Project Cyclone*, an analog computer aimed at studying the missile dynamics. *Project Typhoon* followed soon after in 1952 developed by RCA consisted of 4000 electron tubes and used 100 dials and 6000 plug-in connectors.

Analog computers continued to be in use until the 1980s of last century. Prototypes of hybrid (digital + analog) computers are studied also in present times, and some researchers believe that these will represent the future evolution of automated computing. We will come back to this point in the last chapter.

Among the major advantages of analog computers, there is certainly the speed of execution and the real-time character of the computation. Among the disadvantages, there are the limited precision and their lack of generality in application.

9.3 Electronic Analog Computers

Electronic analog computers are the most common and most versatile form of analog computers and have been used since the 1940s of last century. Their principle of functioning is based on the indirect analogy built by using electronic components. In addition to the already mentioned inductance, capacitance and resistance, a most used component is the *operational amplifier* (op amp). This is an analog integrated device that is operated as a voltage amplifier with two differential inputs and one output, characterized by very high gain. Its use in analog computers is mainly due to the possibility to perform additions and also integrations, making it an ideal building block for analog computation.

In Fig. 9.3, we present the adder configuration. Here, the voltage at the output of the op amp V_o is given by:

$$V_o = \frac{\mathrm{R}}{\mathrm{R}_1} V_1 + \frac{\mathrm{R}}{\mathrm{R}_2} V_2 + \frac{\mathrm{R}}{\mathrm{R}_3} V_3$$

Clearly, if we have just one input, this can also be used as a multiplication by a constant (*amplification* effect).

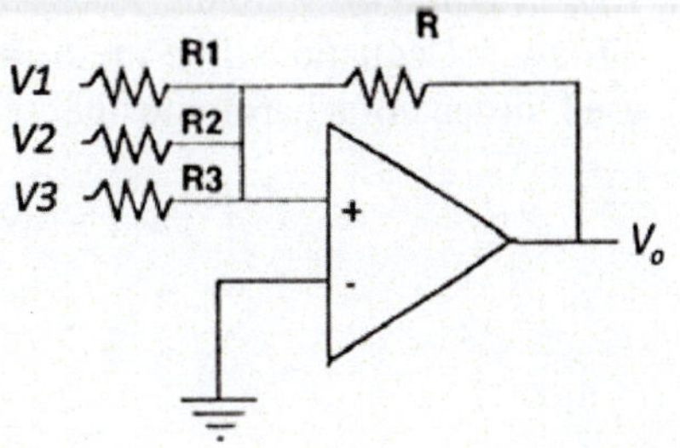

Fig. 9.3 Op amp in adder configuration. The voltage at the input is added, each with an independent coefficient

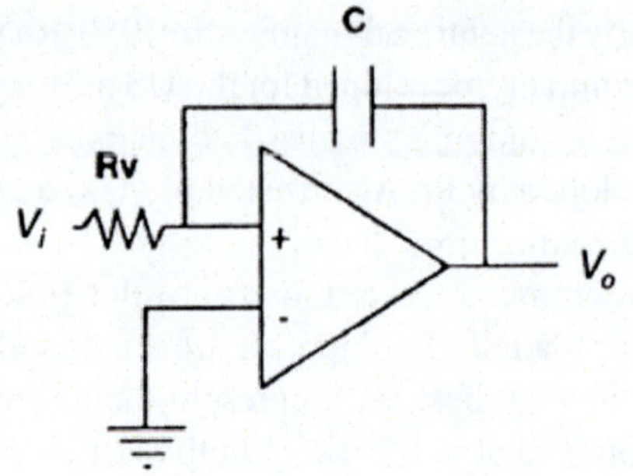

Fig. 9.4 Op amp in integrator configuration

In Fig. 9.4, we have the integrator configuration. Here, the voltage at the output of the op amp V_o is given by:

$$V_o = \frac{1}{\mathrm{R_V C}} \int V_i(t)\, dt$$

In order to understand how do we need to combine these elements, let us consider the simple case where we want to compute the solution of the following first-order linear differential equation:

$$b\dot{x} + cx = y(t)$$

We write it as:

$$\dot{x} = -\frac{c}{b}x + \frac{1}{b}y(t)$$

We use a method that was initially developed by Kelvin (see Fig. 9.5). We draw a logic circuit starting from a point where we supposed to have the highest derivative $\dot{x}(t)$ (indicated by the arrow in Fig. 9.5). If we move down the line (clockwise), once we have integrated, we have x(t). We multiply this term by a constant of value $-\frac{c}{b}$, and thus, we have one of the two rightmost terms in our equation. The other term, $y(t)$, should come as an input from an outside line. Now, we must add the two terms and impose that the result is equal to our starting point $\dot{x}(t)$ by joining the lines.

By using an electronic analog multiplier, a device that provides at the output a voltage proportional to the product of two voltage inputs, we can also treat nonlinear differential equations like the following, where a quartic double-well potential is used, instead of a harmonic one.

$$\ddot{x} = \mathrm{a}x - \mathrm{b}x^3 - \mathrm{c}\dot{x} + F_{ext}$$

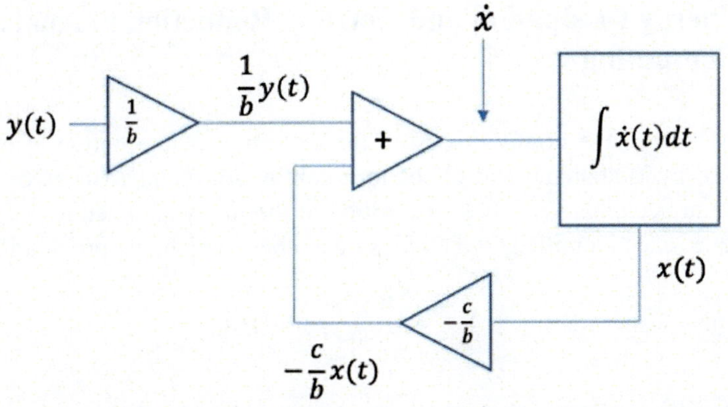

Fig. 9.5 Logic scheme for a circuit that solves $\dot{x} = -\frac{c}{b}x + \frac{1}{b}y(t)$

In the 1980s of last century, when the phenomenon of *stochastic resonance* [2] was discovered, analog computers were built and used to study stochastic differential equations by feeding random voltages generated with a noise generator into an electronic circuit like the one in Fig. 9.6. The integration of such nonlinear equations was by far faster than with digital computers, then. The author of this book did his PhD research on *stochastic resonance*, with the help of these analog computers.

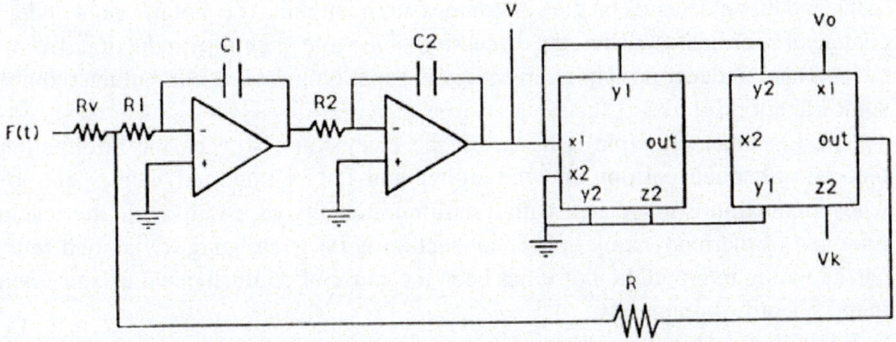

$$\ddot{V} = -\frac{1}{R_1C_1}\dot{V} + \frac{V_q}{10RC_1R_2C_2}V - \frac{1}{100RC_1R_2C_2}V^3 + \frac{1}{R_vC_1R_2C_2}V_i(t)$$

Fig. 9.6 (Top) Electronic analog computer circuit for solving a nonlinear stochastic differential equation (bottom). The two square boxes represent two analog multipliers aimed at generating the cubit force term. F(t) represented the sum of a stochastic and periodic force, as it is typical in stochastic resonance applications

9.4 Energy Dissipation and Entropy Reduction in Analog Computing

In Chap. 4, we have briefly discussed the extension of the notion of entropy, as introduced by Boltzmann and Gibbs to a continuum of microstates by Shannon. There we noticed that W. Gibbs's definition of thermodynamic entropy for discrete microstates, each identified by the index i and characterized by probability p_i, was:

$$S = -k_B \sum_i p_i \, ln(p_i)$$

The available volume in the phase space is measured by the number of available microstates; thus, the sum is taken over all possible microstates; each of them is weighted by the probability that our system actually realizes such a microstate. The extension proposed by C. Shannon has the following form:

$$S = -k_B \int_{-\infty}^{+\infty} p(x) \, ln(p(x))dx \tag{9.1}$$

As we noticed, this is sometimes called *differential entropy* in the literature [3] and, to some extent, does not coincide exactly with Boltzmann–Gibbs entropy for the discrete microstates. The most relevant differences are that it may assume infinite value and that it depends on the system measurement units (i.e. not invariant under a change of coordinates). Thus, the discussion of the role of entropy reduction (like we have in the Landauer principle), in energy dissipation in analog computing, requires some attention [4].

The Landauer principle is based on the connection between the information-theoretic Shannon entropy and the Boltzmann–Gibbs (thermodynamic) entropy, whose reduction is associated with a minimum energy expenditure via the second principle of thermodynamics. This connection is far from being trivial, and in the past, a wrong interpretation of it has been the cause of confusion and disagreement in the scientific community.

For a digital computer, where we have a discrete, finite number of logic states, S_i ($i = 1, \ldots, M$), this connection is based on the identification of a set of M different logic states that are associated with a partition of the configurational space Ω of the computing system. Each element of this partition is identified as Ω_i. When the contribution of the *conditional entropy* that represents the contribution of the different microstates inside each Ω_i is zero, the Shannon entropy is equal to the thermodynamic entropy generated when the reset is performed. In this case, the variation of the information (Shannon) entropy ΔS_S between the final state, in which the system is in a specific state with probability 1 and has thus $S_S = 0$, and the initial state in which the systems can be in any one of the possible states S_i with probability p_i is given by

$$\Delta S = k_B \sum_i p_i \; ln(p_i)$$

For an analog computing system, before the connection between the (differential) (Shannon) entropy and the thermodynamic entropy is made, an appropriate regularization that adapts the dimensional character of the relevant degree of freedom to the dimensionless quantity considered in the probability density $p(x)$ is required. To cure this problem, it has been proposed [5] to modify Eq. (9.1) by introducing an invariant factor $\mathrm{p_k(M)}$ that represents the density of the discrete distribution that gives $p(x)$ in the continuum limit:

$$S = -k_B \int_{x \in \mathrm{M}} p(x) \; ln\left(\frac{p(x)}{\mathrm{p_k(M)}}\right) dx$$

As we will see in the next chapter, the physical interpretation of this invariant factor is to provide a minimum, nonzero volume in the phase space. Its existence is justified in the view of the indetermination principle proposed by Heisenberg, in the framework of quantum dynamics.

More, interesting, developments await us in the next chapter, and surprises will not lack.

References

1. https://wikimedia.org/wiki/File:NAMA_Machine_d%27Anticyth%C3%A8re_1.jpg
2. L. Gammaitoni, P. Hänggi, P. Jung, F. Marchesoni, Stochastic resonance. Rev. Mod. Phys. **70**(1), 223 (1998)
3. M. Thomas, J.A. Thomas, *Elements of Information Theory.* (Wiley, New York, 1991), ISBN 0-471-06259-6
4. M. Cristina Diamantini, L. Gammaitoni, C.A. Trugenberger, Landauer bound for analog computing systems. Phy. Rev. E **94**, 012139 (2016)
5. E.T. Jaynes, *Statistical Physics, Brandeis Lectures in Theoretical Physics*, vol. 3 (W. A. Benjamin Inc., New York, NY, 1963), pp. 181–218

Chapter 10
Computing in the Quantum Regime

I do not like it, and I am sorry I ever had anything to do with it.
Erwin Schrödinger

Abstract In this chapter, we discuss the role of quantum physics in the functioning of computing devices. Quantum mechanics, the first branch of quantum physics, was introduced at the very beginning of the twentieth century, when computers had just begun their long journey. Quantum mechanics, in many aspects, differs substantially from classical mechanics that is the basic physics body of knowledge that we used in previous chapters. Here, we will see how this "being different" has implications for performing computational tasks for those devices where quantum effects cannot or do not want to be ignored.

10.1 How is Quantum Computing Different from Classical Computing

In Chap. 2, when we introduced digital computing, we set the minimum conditions for a computing device as:

(1) a *physical system* capable of assuming two different *physical states*
(2) a way to induce state changes in this physical system (typically a force)

where a *physical state* was a given set of measurable quantities, whose value can be used to distinguish unambiguously two different outcomes. When it comes to deal with very small objects, of the order of a nanometer and below, these conditions still hold but the concept of physical state and the role of the observable quantities that define it need some revision. When the size of our computing device gets smaller and smaller, we enter what the physicists call the *quantum realm*, i.e. a different set of rules for describing the behaviour of nature.

L. Gammaitoni, *The Physics of Computing*,
https://doi.org/10.1007/978-3-030-87108-6_10

We do not have the room here to deal in detail with the *quantum realm* but, just to give you a glimpse of what we are talking about, we will briefly discuss few phenomena that well account for the different behaviours of such systems.

The first phenomenon is the *wave–particle dualism.* The dynamics of physical systems in the quantum realm cannot be described in terms of the Newton equation that was invented to explain the way material points move, nor it can be described in terms of a wave equation, as proposed by Jean le Rond d'Alembert (1717–1783) in 1746 to explain the motion of wave-like perturbations in a string. We need a different equation of motion that was initially proposed by Erwin Schrödinger (1887–1961) in 1925. The Schrödinger equation, as it is called today, is a linear partial differential equation that describes the evolution of the *wave function* that is the complex-valued time and space function representing a physical state. As a consequence of the many experiments made in the last 100 years, we have learned that a physical state does not behave like a particle, nor like a wave, but it has features that belong to both of these categories. An interesting feature, for example, is that an electron (to say a small object whose physical state is in the *quantum realm*) can *be* at the same time in two different positions. Another interesting feature is that it is not possible to determine the position and the momentum of an electron with arbitrary precision. Every attempt to measure these conjugated quantities will result in an uncertainty quantified by the, already mentioned, Heisenberg indetermination principle.

The second peculiar phenomenon is the so-called *entanglement.* It happens that a group of small systems (e.g. electrons) that have been in close proximity develop physical properties, as, for example, the value of angular momentum, that keep being strictly correlated, also if time is passed and they are separated at large distance from each other. This property is of great interest for applications also in the field of cryptography and communications.

The third quite relevant feature is that the very concept of time in the *quantum realm* assumes a different perspective. To see what is different, let us consider a free electron that at time $t = 0$ is in a given position $x(0)$, affected by an uncertainty $\Delta x(0)$ and a velocity $v(0)$ affected by uncertainty $\Delta v(0)$. If we were in the *classical realm,* the uncertainty of the electron position would evolve according to:

$$\Delta x(t) = \Delta x(0) + \Delta v(0)t$$

In the *quantum realm,* instead, the time evolution of the position uncertainty is:

$$\Delta x(t) = \sqrt{\Delta x(0)^2 + \Delta v(0)^2 t^2}$$

with

$$\Delta x(0)\Delta p(0) \geq \frac{\hbar}{2}$$

due to the Heisenberg indetermination principle. Being $\Delta p(0) = m\Delta v(0)$, we have:

$$\Delta x(t) \geq \sqrt{\Delta x(0)^2 + \left(\frac{\hbar}{2m\,\Delta x(0)}\right)^2 t^2}$$

As we can see, the uncertainty in the position will keep growing with time. While this phenomenon can be in principle avoided in the *classical realm*, by making arbitrarily small $\Delta v(0)$, In the *quantum realm*, this is not possible and, no matter how small we start with $\Delta x(0)$, $\Delta x(t)$ will keep growing. Another way to look at this behaviour is that, while in the *classical ream* time is an external parameter and immobility (i.e. $\Delta x(t) = \Delta x(0)$) is not in principle forbidden, in the *quantum realm* immobility is forbidden and time is unavoidable, being intrinsically connected with the physical system itself.

Based on these examples, we can conclude that quantum computation is different from the classical computation due to the different features of the physical state that is considered to represent the computing device. In the following paragraph, we will briefly review how these new features can be exploited to do computation.

10.2 Quantum Computation and Energy Dissipation

Presently, with *quantum computation* it is intended that the exploitation of the properties of quantum states is to do digital computing. The devices are employed to do such a task to show the features of the quantum realm, and for these reasons, they are usually very small systems that belong to the atomic level of the matter.

The practice of quantum computing is quite recent and dates back to the 1970s and 1980s of last century. Paul Benioff, at that time at Argonne National Laboratory, proposed [1] in 1980 a quantum mechanical model of the Turing machine. Other important contribution to the topics was also provided by Richard Feynman and Yuri Manin.

Since then, different approaches to quantum computing were proposed. Two of the most popular are: the *quantum circuit* model and the *adiabatic* quantum computer.

The quantum *circuit model* is based on the classical digital computation scheme, where logic gates are employed in order to build networks aimed at performing a given algorithm on binary digits. Here, we have quantum logic gates and quantum binary digits (*qbits*). A quantum gate is a logic device that exploits the features of the quantum realm, by manipulating digital quantities. Such quantities are input as digits and read at the output as digits but inside the network of quantum gates are represented as binary quantum states, called *qbit*. In a binary quantum device, we have two physical states that can be associated with the logic "0" and "1". A *qbit* can be in the logic "0" or in the logic "1", or in a superposition of the "0" and "1" logic states. At the end of the computation, due to the interaction with the measurement system, the computation output, however, will always be "0" or "1" exclusively.

Although the theoretical aspects of the quantum circuit model have been studied in detail and the associated phenomena well understood, there are a number of significant technical difficulties in making a large-scale (i.e. able to treat a large number of *qbits*) quantum computer. The most relevant critical aspects are associated with the difficulty to let the quantum state to evolve as desired without unwanted interaction with other systems. These unwanted interactions define the so-called *decoherence time*; that is, the maximum time during which a gate logic operation has to be performed. The more isolated the system is, the larger is the *decoherence time.*

A different, less popular, computing scheme is represented by *adiabatic quantum computing*. Here, the state evolution that characterizes the computing activity is carried on by keeping the system always in the ground state. Technically, the computing starts with identifying a proper Hamiltonian that represents the final state, i.e. the solution of the problem to be computed. Then, a physical system is prepared in an initial state represented by a simpler Hamiltonian and slowly let evolve into the final Hamiltonian. The goal is to keep the system always in the ground state, and this is simpler if the energy gap between the ground state and the first excited state is large enough. As a rule of thumb, the evolution time has to be of the order of $1/g^2$, with g the energy gap.

The *adiabatic quantum computing* is a potentially valid strategy to avoid the interference of the outside world with the quantum state, during the computation. As a matter of fact, if the system is in the ground state, any interaction cannot make it to relax into a lower energy state (thus dissipating energy) and it will keep the coherence as long as necessary.

As we have discussed for the classic realm, also in the quantum realm, dissipation of energy is a relevant issue for computing devices. The *quantum dissipation* is the quantum analog of the classical dissipation discussed in Chap. 3. As before, the typical approach to deal with this subject is to identify few relevant dofs that represent the system evolution and to put all the others dofs into a thermal bath that exchanges energy with the system.

One of the simpler models for the interaction with the thermal bath is the so-called Caldeira–Leggett model, proposed by Amir Caldeira and Anthony J. Leggett, in 1981. The thermal bath is made by a very large number of independent quantum harmonic oscillators, following a suggestion advanced by Richard Feynman twenty years earlier, that are linearly coupled to the system dynamics. For a one-dof system, the Hamiltonian of the system coupled to the bath is:

$$\mathcal{H} = \frac{p_x^2}{2m} + V(x) + \sum_{i=1}^{N}\left(\frac{p_i^2}{2m_i} + \frac{1}{2}m_i\omega_i^2 q_i^2\right) - x\sum_{i=1}^{N} C_i q_i + x^2 \sum_{i=1}^{N} \frac{C_i^2}{2m_i\omega_i^2}$$

The first and the second terms correspond to the Hamiltonian of a particle of momentum p_x and mass m subjected to a potential $V(x)$. The third term represents the energy of the thermal bath made by N (N may be infinite) harmonic oscillator modes (like bosonic particles), with position q, momentum p and frequency ω. The next term represents the coupling between the particle and the bath, where the C is the coefficient that accounts for the strength of the coupling with a specific mode. Finally, the last term is necessary in order to assure that the model is translationally invariant.

The idea behind this approach to quantum dissipation is to find a proper Euler–Lagrange equation that describes the system coupled to the bath and shows that, under appropriate conditions, the classical Langevin equation (typical of the Brownian motion) can be derived by it. Once this is done, the quantum dynamics of the Hamiltonian can be studied through the usual Schrödinger equation formalism.

10.3 Quantum Binary Switches

As we have seen in Chap. 2, the binary switches are the basic element for building a digital computing device. In the classical realm, we have derived a mathematical model, in the presence of dissipation and fluctuations, to deal with the switch dynamics.

In the quantum realm, we can use the Caldeira–Leggett approach for studying a dissipative two-state system that can represent the quantum analog of the binary switch. In this case, we write the Hamiltonian function for a two-state system, by focusing on the dynamics of the spin ½, for example, and thus, we change from the continuous dynamics of the position operator to the discrete dynamics of the two-state spin. This is sometimes interpreted as the dissipative model for the quantum binary switch evolution. The two-state Hamiltonian can be written as:

$$\mathcal{H} = \frac{\sigma_x}{2}\Delta + \sum_{i=1}^{N}\left(\frac{p_i^2}{2m_i} + \frac{1}{2}m_i\omega_i^2 q_i^2\right) + \frac{\sigma_z}{2}\sum_{i=1}^{N} C_i q_i$$

where σ_x and σ_z are the two-by-two Pauli matrices and Δ is the probability of transition between the two states. The other terms are as in the previous equation. Note that no additional term for invariancy is necessary here.

Another example of two-state dynamics is represented by the so-called Jaynes–Cummings model used in quantum optics. It was developed by Edwin Jaynes (1922–1998) and Fred Cummings (1931–2019) in the early 1960s of last century, to describe the dynamics of a two-level atom interacting with the electromagnetic (quantized) field of an optical cavity.

10.4 Quantum Limit in Computing Speed

How fast a computing device can perform its operations? This is certainly an interesting question, by the moment that increasing the computation speed has been the main goal of the last forty years in the computer industry. This race for speed has been sustained by the technological/economical curve known as Moore's law, after Gordon Moore, co-founder of Intel Corporation. With the doubling of the transistor density in modern chips every two years or so, also the computational speed has grown at an exponential pace, with the goal of reaching exascale computers (10^{18} operations per second) as the present state of the art in desirable objectives.

Question is: is there any fundamental law of physics that limits how fast a computer can go? According to what we discussed in chapter two, the elementary act of computing is a switch from a given physical state (e.g. S_0) to another physical state (e.g. S_1). In the *classical realm*, as we have seen, the speed in the switch operation is associated with the amount of energy that is dissipated due to the unavoidable presence of friction. Depending on the dissipative model, the increased speed carries an increased specific amount of dissipated energy, but no fundamental limit applies. The situation in the *quantum realm* is slightly different due to the Heisenberg indetermination principle.

Let us start by considering two quantum states that differ in energy by a quantity ΔE. We can assume a ground state with energy E_0 and an excited state with energy $E_1 = E_0 + \Delta E$.

The two physical states S_0 and S_1 can be written in terms of the eigenstates of the energy, $|E_0\rangle$ and $|E_1\rangle$ as follows:

$$|0\rangle = \frac{1}{\sqrt{2}}(|E_0\rangle + |E_1\rangle) \text{ and } |1\rangle = \frac{1}{\sqrt{2}}(|E_0\rangle + |E_1\rangle)$$

The state change can be obtained by applying a switch operator that brings $|0\rangle \rightarrow |1\rangle$ and $|1\rangle \rightarrow |0\rangle$. The two states are orthogonal to each other and separated by an energy distance ΔE. According to an interpretation [2] of the time–energy relation of the indetermination principle, proposed by Margolus and Levitin [3] it should require a minimum time Δt to evolve from one state to the other, such that:

$$\Delta E \Delta t \geq \frac{\hbar}{2}$$

Based on this interpretation, Seth Lloyd, at MIT, has proposed that the maximum computing velocity has to be set as the inverse of the minimum switch time

$$v_{max} \leq \frac{1}{\Delta t} = \frac{2E}{\hbar}$$

where $E = \Delta E$, being the average energy to be invested in the switch event. According to this interpretation, the computing speed is only limited by the amount of work that one is willing to do, to perform the switch. Although this idea is quite suggestive, it is based on an interpretation of the indetermination principle that has been long debated. This is a field where we could benefit from some good new experiment that sets the final score on the maximum achievable speed. Any volunteer?

10.5 Quantum Limit to Entropy Reduction in Analog Computing Systems

We go briefly back to the *classical realm*, for discussing some interesting consequences of the extension of the Shannon entropy for analog computing systems. We discuss this aspect here, because such consequences are connected with the Heisenberg indetermination principle (yes! The old, good Heisenberg again). As we have seen in Sect. 9.4, the extension of Shannon entropy to continuous systems has to be taken with some attention, due to the regularization issues that it requires. In the following, we will focus on this aspect and present a reasoning [4] that will show how, by taking into consideration the indetermination principle, the invariant factor $p_k(M)$ introduced in Sect. 9.4 arises naturally.

We focus our attention on a popular quantum system, the so-called ferromagnetic Heisenberg model, where a spin system can take all values on a sphere of unit radius. We have, thus, a continuum, infinite number, instead of a discrete, finite number, of possible values for the spin vertical (z) component.

The three-dimensional ferromagnetic Heisenberg model for N spins, subjected to an external magnetic field H in the z direction, is described by the Hamiltonian

$$\mathcal{H} = -\frac{J}{2} \sum_{<i,j>} s_i s_j - H \sum_i s_i$$

where $J > 0$ and $|s^2| = 1$. We will consider the mean-field approximation. It is known that the spin system undergoes a phase transition when the temperature T becomes lower than a given critical temperature T_C. This phase transition is characterized by an order parameter, the mean magnetization m, increasing from the value 0 above the critical temperature (disordered phase) to 1 at zero temperature (perfectly ordered phase). Lowering the temperature, thus, is similar to a reset process, where a continuous spin is reset to a standard value. The overall magnetization m accounts for average magnetic field generated by all other spins and plays also the role of a quantifier for the error probability in the reset operation.

The objective is to compute the entropy variation during the reset process and to compare it with the change in the Shannon (differential) entropy. In order to simplify the analysis, we are going to compute the quantity

$$-\frac{\Delta S}{k_B N} = \frac{S(T_C) - S(0)}{k_B N}$$

that expresses the change in entropy per spin, per unitary k_B. Details of the calculation are available in the cited paper. Here, we report the result in the two relevant limits:

$m \to 0$ we have:

$$\frac{S(T_C)}{k_B N} = \log(4\pi)$$

$m \to 1$ we have:

$$\frac{S(T \to 0)}{k_B N} \to -\infty$$

As we anticipated, the negative divergence of the entropy is a problem common to the continuous extension of Gibbs entropy. In order to overcome this problem, we consider the classical Heisenberg model as the proper classical limit of its quantum counterpart. A negative divergent entropy at $T = 0$ would imply an infinite amount of work necessary to align the spin (perfect reset operation). In the mean-field approximation, the quantum ferromagnetic Heisenberg model is the model describing a system of quantum, *non-interacting* spins, with a simplified Hamiltonian and an effective magnetic field H_{eff}:

$$\mathcal{H} = -H_{\text{eff}} \sum_i s_i$$

In this case, it can be demonstrated that the entropy, in the $m \to 1$ limit, has the correct behaviour:

$$\frac{S(T \to 0)}{k_B N} \to 0$$

Now, in order to take the classical limit of the quantum Heisenberg model, the $(2s + 1)$ values of the quantum spin need to be properly distributed on the classical sphere of area 4π. In doing this, we need to introduce a density Δ that represents the minimum area in the unitary sphere occupied by our spin, allowed by the Heisenberg principle. By carrying on the calculation, we obtain:

$m \to 0$ we have:

$$\frac{S(T_C)}{k_B N} = \log(c)$$

$m \to 1$ we have:

$$\frac{S(T \to 0)}{k_B N} = \log \frac{\hbar}{2}$$

In this perspective, it is easy to understand that the negative divergence in the classical entropy is due to the fact that the classical probability distribution can be concentrated in regions smaller than the minimum area. Δ is in fact the invariant factor $p_k(M)$. The presence of this factor is not due to the specific model we considered, but is a general consequence of the fact that the continuous Shannon entropy must be regularized in order to make it invariant upon a change of coordinates.

Finally, we are in position to compute the change in entropy as a consequence of a reset in an analog computing system:

$$-\frac{\Delta S}{k_B N} = \frac{S(T_C) - S(0)}{k_B N} = \log(4\pi) - \log \frac{\hbar}{2} = \log \frac{8\pi}{\hbar}$$

This is the analog generalization of the Landauer bond where the entropy reduction during the reset process is given by the available configuration volume (the area 4π in this case) measured in units of the minimum quantum of configuration volume Δ.

There is an interesting consequence that becomes apparent if we consider the quantity of *information* that can be encoded in a physical system: even if we start with continuous, analog information, due to the existence of a minimum phase space volume, only a finite countable amount of information can be encoded [5].

References

1. P. Benioff, The computer as a physical system: a microscopic quantum mechanical Hamiltonian model of computers as represented by turing machines. J. Stat. Phys. **22**, 563 (1980)
2. S. Lloyd, Ultimate physical limits to computation. Nature **406**, 1047 (2000)
3. N. Margolus, L.B. Levitin, The maximum speed of dynamical evolution. Physica D **120**, 188–195 (1998)
4. M. Cristina Diamantini, L. Gammaitoni, C.A. Trugenberger, Landauer bound for analog computing systems. Phys Rev E **94**, 012139 (2016)
5. J.D. Bekenstein, Phys. Rev. D **30**, 1669 (1984)

Chapter 11
The Future of Computing and Society

The little motes of dust swirling in a shaft of sunlight in a dark room.
Lucretius, (II.114-124)

Abstract We devote this last chapter of the book "the physics of computing" to reflections on the role of computing in present and future society. Still, the physics rules what it is and will be possible to do with computers in the foreseeable future. In the following, we discuss the role of artificial intelligence, as realized through computing devices and the great availability of digital data. We critically discuss the role of data analytics in an increasingly interconnected society and, ultimately, take a look at our technological future, where analog and digital computers will most likely end up merging into a form of hybrid computer that operates in the presence of significant random fluctuations.

11.1 Reflections on the Computing Era

There is little doubt that the introduction of automatic computing has been of enormous importance in our society. Our present culture has been shaped, at least for the last 70 years, by computers and their technology. Not a day passes that we are not reminded that we live in the age of information.

It is very important for the future of our civilization that we do not give for granted the present condition nor that we overlook at the possible risks associated with such a great development. The functioning of the information society and the risks associated with it have been the subject of many popular books and papers. Among them, the article written by Shoshana Zuboff (at the Harvard Business School) and published in the New York Times on 19 January 2021 is probably the most passionate and preoccupied. The author is concerned with what she calls the growing danger of *surveillance capitalism.*

Zuboff's is just one of the many alarmed cries about a growing risk of manipulation of democratic rules, thanks to the pervasiveness of digital technologies. The

L. Gammaitoni, *The Physics of Computing*,
https://doi.org/10.1007/978-3-030-87108-6_11

scandal created by the British company Cambridge Analytica, accused of having manipulated information disseminated by the media for electoral purposes during the US campaigns and the Brexit referendum, has brought the problem to everyone's attention. If that were not enough, the uncontrolled spread of fake news that has made the COVID-19 pandemic even more lethal reminds us that the danger has not been overcome but, on the contrary, is thriving and the idea is spreading that it is necessary to promote and make grow a *data culture*. Prestigious universities compete to offer *data science* courses in order to train new *data scientists*, as normal statistics courses were not enough. *Data science* courses should certainly be instituted, but not in information engineering. Rather, they should be promoted for political science degrees, so that the consequences of this recent technology on the functioning of our society can be studied in depth.

11.2 From the Introduction of Digital Technology to Big Data

The risk to our civilization exists because data exist, and data exist because digital technology exists.

Digital technology was born in a time where noise was an enemy and electronics was still in the cradle. In the early 1940s, sending a message was not an easy task and the quality of the one received often differed greatly from the original, due to the unwanted action of noise. The natural solution was to reduce the variability of the signal, due to fluctuations, by imposing a finite number of amplitude levels.

It is important to remember that such a reduction, called digitalization, i.e. the action aimed at transforming a continuous (analog) signal into a discrete (digital) one, is an operation that is not free of errors. It mainly consists of two different procedures: sampling and quantization. Sampling is the time periodic measurement of the signal. If this procedure is accomplished properly, according to the work done by the Swedish-American electronic engineer Harry Nyquist (1889–1976), no detail of the original signal is lost. Quantization, on the other hand, is the transformation of a continuous smooth signal into a stepwise one. This operation cannot be accomplished error-free, and no matter how carefully we perform it, there will always be an error. This implies that the digitalized signal will always be different from the original one.

Notwithstanding this limitation, scientist has used digitalization to build an enormous societal progress based on the extensive use of digital tools. Undoubtedly, the most popular of them is the digital computer. In addition, we have had digital communication systems, like digital phones, digital radios and digital TVs, and also digital industrial machines and most of the present robots that operate in the manufacture sector. Going digital has become the new mantra, and most countries have invested large amount of money just to promote the so-called digital transformation, i.e. the massive introduction of these tools into our daily life.

In order to perform their tasks, these digital tools need input data that are captured from the physical world through digitalization procedures, performed by special devices called analog-to-digital converters (ADCs). Data production in present-day society has become enormous, and the large quantity of data available has triggered some concern about its maintenance, ownership, control and use.

The impact of the digital technology has affected not only the way we work and communicate but also our understanding of the world around us. As we have discussed in Chap. 7, it was Ralf Landauer that, building on the work of Shannon and John von Neumann, firstly advanced the idea that the amount of information stored in a computer memory could play a role on the physics of computing itself, being interpretable as part of the entropy of the computer. *Information is physical* was the motto that popularized this idea. Although that statement was long debated in the physics community and I, myself, among others, worked to show that his conclusion was wrong [1], the idea that the application of the information theory paradigm, i.e. the digital perspective underlying it, was a potentially useful key to interpret the fundamental laws of the universe flourished. The most famous representative of this idea was probably John Archibald Wheeler (1911–2008), the physicist that in 1967 invented the name "black hole" for the mysterious celestial objects, recently observed. Wheeler, during a conference held at the Santa Fe Institute, in the spring of 1989 proposed the expression "It from Bit", to exemplify the process that we go through when we deal with the reality of things. In Wheeler's idea, expressed also in "Information, Physics, Quantum: The Search for Links" [2] our experience is the result of binary decisions (thus expressed by bits: 0 or 1) that we take during each observation. The underlying fabric of the universe is based on these decisions, and its nature is intrinsically digital.

More recently, a similar point of view has been advanced by Seth Lloyd that published a popular book entitled "Programming the universe" (2006) where he discusses this matter.

My point of view is that information theory, although a nice and useful theoretical framework for dealing with computer programming, is actually not adding much to our understanding of physics. In some cases, it may offer an interesting perspective but still, fundamental laws of physics need a more general point of view that those based on the idea that everything, at the end, is a mere computation act. Is this the end of it? Not really. Recent progresses on artificial intelligence (AI) studies found a profitable application for the large amount of data produced by our digital tools. Deep learning (DL) algorithms, very powerful statistical classification schemes, proved useful in image and speech recognition. Although progresses in other aspects of AI are still lacking behind, some scientists are applying DL to the task of guessing new laws of physics and new science schemes in general.

11.3 Big Data and AI for the Creation of New Knowledge

The successful approach of DL technologies to image recognition has soon prompted a greater ambition: the substantial overcome of the old-fashioned scientific method. Chris Anderson, editor-in-chief of the American magazine Wired, in 2008 wrote a piece entitled "The end of the theory: the deluge of data makes the scientific method obsolete", where he claims that the great availability of "experimental" data let us overcome the traditional scientific approach based on the creation of physical models (meaning theories created by the mind of scientists and based on mathematical relationships) and fosters a new approach based, instead, on the massive use of AI techniques. The idea is to skip the search phase of an existing but hidden model and go directly to the identification of patterns in the data. In the words of Peter Norvig, director of research at Google and AI expert: "all models are wrong and more and more often you can find the solution without using them". The message is clear: dear scientist, please stop trying to figure out how the world works, computers will find for us the laws of nature that will not necessarily take the form of intelligible mathematical relationships, but will work just as well.

The idea of looking for something, which we do not know if there is and exactly how it is, is not really new and has been at the heart of the work of physicists and mathematicians, who dealt with the so-called CAOS theory, since the eighties of the last century, among them the Dutchman Floris Takens and the Belgian-born David Ruelle. At that time, they were trying to find a way to decide if, given a set of apparently random data, these were actually random or were the product of dynamical system undergoing a chaotic evolution.

Nowadays, however, we are one step ahead compared to the program of Ruelle and Takens: as a matter of facts, the idea that circulates now is to skip the search phase of an existing but hidden model and go directly to the identification of patterns in the data. Is this a possible route to new discoveries? Are we really convinced that the availability of AI programs and large quantities of empirical data will make the traditional way of doing science, based on the work of scientists to create logical–mathematical models of the world, useless?

There are several reasons why the "big data" program, at least in its most radical version, cannot work. Some of these reasons have been discussed by Hykel Hosni and Angelo Vulpiani, in a paper entitled "Forecasting in light of big data" and published in the specialized journal "Philosophy & Technology" [3].

Besides these purely technical reasons, there is one that has deeper roots than the others and we want to talk about this here [4]. To better understand what it consists of, we will use a literary aid: the famous story by Jorge Luis Borges entitled *The library of Babel*. In this story, he imagines the existence of a vast library composed of a huge number of hexagonal rooms, all the same, connected by corridors. As Borges explains: *Each wall of each hexagon corresponds to five shelves; each shelf contains thirty-two books of uniform format; each book is four hundred and ten pages; each page, of forty lines; each a row, of forty letters of black colour.* As will soon be discovered in the story, each book is composed of a random sequence of symbols (22 letters, plus the comma, the point and the space). The library contains all the possible combinations of these symbols and, thus, contains all the writable books that satisfy the conditions of length above expressed. To use Borges' poetic language, *the library's books describe everything: the meticulous history of the future, the autobiographies of the archangels, the faithful catalogue of the Library, thousands and thousands of false catalogues, the demonstration of the falsity of these catalogues, the demonstration of the falsity of the authentic catalogue, the Gnostic gospel of Basilides, the commentary of this gospel, the commentary of this gospel, the truthful account of your death, the translation of each book in all languages, the interpolations of each book in all the books.*

The problem facing the visitor of the library is obviously to decipher the books because, as you can imagine, a book taken at chance from the shelf appears as a sequence of symbols randomly assembled to compose meaningless words. No sense for us, but perhaps expressing the story of our life or a prophecy or even the final equation of Physics, in another unknown and mysterious language.

A book from the library of Babel, if you think about it, looks just like a series of data gathered from a possible experiment, provided that you assign a number to each symbol. It is apparently a sequence of random numbers but you never know if there is some hidden promising signal, within the ocean of noise. Therefore, looking for a new signal in the experimental data series would not be very different from looking for a sensible and interesting expression, in one of the library's books. And here comes the beauty: once this analogy is established (a book as a bunch of data), the *Library of Babel* really looks like the big data paradise. It contains all the information of potential interest to us, and the problem is "just" extracting this information. Before you run out to apply for the next university course on big data, however, I have the obligation to warn you: things can be a bit more complicated than expected.

To come to our aid is an Italian mathematician, Lucio Lombardo Radice, who in 1981 wrote a nice booklet, entitled "L'Infinito", of great interest for our problem. Lombardo Radice explains that, even if the number of books in the Library of Babel is very large but still finite, the number of meanings that can be attributed to the content of those books is not. Technically, this is called "Richard's paradox" and is part of a family of results that led the logical mathematician Gödel to formulate his famous incompleteness theorems. An infinite number of meanings potentially correspond to a finite string of characters, or in our context, a certain set of data constitutes the potential answer to an infinite number of physical questions. Without knowing the question, the answer risks being meaningless.

To better understand this point, let us see an example. Suppose in one of the books, in the middle of senseless strings of characters, there finally appears a sequence written in a familiar language: "while the music goes, Alice and Bob exchange secure messages through their entangled spins". What is the meaning of it? The two words "entangled spins" indicate two very different things depending on whether you interpret them using a nineteenth-century English dictionary or a late twentieth century one. In the first case, the phrase would indicate the confidences that two lovers exchange, perhaps whispering mouth to ear, during a waltz tour. In the second case, on the other hand, it would take on a completely different meaning since in the twentieth-century physics, thanks to the creation of quantum mechanics, has emerged the concept of "entangled spins" which indicates two microscopic physical systems that have related properties and can be used for encrypted communications. Thus, to a contemporary reader, the phrase would sound like: while the music goes, Alice and Bob exchange safe messages through quantum cryptography. Which one of the two interpretations is the correct one? It depends on the application, which depends on the dictionary I use.

Out of metaphor, the construction of the scientific model (the dictionary in our example) is crucial for the interpretation of the experimental data. Without the laborious, complicated and "dirty" work of the scientist who mixes intuition and induction, genius and creativity, and manages to advance hypotheses that will then be denied or confirmed by data, there is no production of true knowledge. To paraphrase the great French scientist Poincarè, we could say that *the scientist must make order: science is made with data as a house is made with bricks, but the accumulation of data is not science any more than a pile of bricks is a house.*

In short, it is really difficult to find something new if you do not know what you are looking for and learning to ask interesting questions is often much more useful than rummaging through abundantly available answers.

11.4 The Surveillance Society Nightmare

We opened this chapter with the potential threat represented by the so-called *surveillance capitalism.* Think for a moment at the following imaginary scenario.

In a fictional, dystopian tale, a small group of powerful and evil men decide to devise a plan to control the planet. They understand soon enough that in order to reach their goal it is necessary to control people, and to better control them, it is necessary to be able to reach them one by one with an effective and pounding communication. So personalized and subtle as to be almost invisible to the mass and, nevertheless, strongly persuasive and convincing. All this, of course, without the knowledge of the people themselves. How to do?

It is a big problem. Initially, an attempt is made to establish ruthless dictatorships but the yearning for freedom is too strong. Then, they try the press, set up several newspapers and control their publishers. But even this path fails, just as radio, TV and the whole telecommunications sector fail. Years pass without the proposed solutions giving acceptable results. When, now discouraged, they are about to abandon their plan, a new idea arrives: give them a new electronic toy. A kind of portable phone that also makes use of images (mostly self-taken photographs and cat footage) and music.

It works great: in a few years, the toy multiplies in increasingly captivating versions and a significant percentage of humanity uses it on a daily base, sharing confidential aspects and intimate news in public. In private, the new toy is able to trace your position, to listen to your voice, to know if you run or walk, if you sleep or are awake. It can monitor your heartbeat and take note of your health. Moreover, all the data that it does collect, you need not to worry: they are safely stored elsewhere, in good hands.

Soon enough, the fun toy realizes, almost for fun, the dream of all the secret services: to have direct and unfiltered access 24/7 to billions of people. Ready and willing to provide information about their religious, political and economic thoughts. Available to be reached individually, to be advised, influenced and, ultimately, directed. And all this is because of a simple and harmless toy, practically child's play.

This is undoubtedly an imaginary plot that has little resemblance to our reality, and obviously, information about us and our habits existed even before the arrival of digital devices and the dusty archives of the secret services have been around since long time. However, the extraordinary ability to aggregate, analyse and use data is a clear consequence of the existence of what we call *digital culture.* Without the huge amounts of data produced every day, especially thanks to the Internet and our use of the software tools that it makes available to us (social networks and communication apps), digital culture would not exist and neither would the looming threat of a *surveillance society*.

11.5 Are Digital Data Inevitable?

In many of the present worried comments circulating about the threat represented by the large availability of personal data, a common factor is the *inevitability* of them. Their omnipresence and their indispensability to ensure the development of knowledge and the quality of life are given for granted. But is not so.

Data are not essential; on the contrary, data are a hindrance and, potentially, the main cause of the growing threat to a fair and a democratic development for our society.

The first step in understanding why data are not essential is to comprehend that data are a technological product: they were not there just a hundred years ago, and most likely, they will not be in a hundred years. Like all technological products, it will happen to them what happened to the texts written on the scrolls that used to populate the Great Library of Alexandria: they will be replaced. Medieval libraries were crowded with manuscripts and hosted just few scrolls. Renaissance libraries, on the other hand, were teeming with printed books (another product of technology) and running short of manuscripts and scrolls. Contemporary libraries have fewer and fewer printed books and more and more digital texts. Libraries of the future will have something we have not invented yet but will inevitably be in short supply of digital texts.

Data, as a technological product, are created through the common activity of each of us, many times a day, without us even noticing. As an example, we are digitizing every time we put an electronic thermometer under the armpit. After a few seconds, a number appears (i.e. data) which informs us about our temperature. Similarly, when we weigh the flour to make a cake, on our electronic scale, we read the weight value on the display: another data. The production of data, and this is the important part of our reasoning, is our decision because, for example, we could choose to use an analog thermometer and then read the temperature value on a graduated scale, where a coloured liquid has grown close to a certain mark. In this way, we would not do any digitization; that is, we would not produce any data, but we would still got the information we were looking for. Equally simple is to use an analog scale where the metal needle of a steelyard oscillates, more and more slowly, to stop near another mark. Finally, are still among us those who remember the old analog phones that allowed us to talk from one end of the world to the other, perhaps adding a background rustle of increasing intensity with the distance of the call, through which information can be communicated without for this to transfer any digital data.

It may be objected that these simple examples all refer to devices of the past, with many defects and few advantages, starting with the slowness, the bulk, the poor accuracy. Again, this is not the case. First of all, because analog instruments are always able to give us a more correct information than digital ones. The reason is that in every digitization we do, we insert an inevitable error. In data analysis courses, university students learn that this is called "quantization error" and no matter how hard you try, you can only reduce the extent of the error but never eliminate it completely. Digital instruments are and always will be affected by an error that analog instruments in principle do not have. The speed of operation, the dimensions and the weight are all attributes that depend on the technology. Analog electronics existed well before digital electronics and continue to work very well.

But, you might argue, computers are digital and who in their right mind would feel like giving up the enormous benefits that computers bring us today? Answer: no one, because we don't have to give up computers. Not only there were analog computers well before digital ones (as we have widely discussed in Chap. 9), but the computers of the future will probably be hybrid computers where digital and analog will coexist, with a large prevalence on the analog functioning over the digital interface. Even the information storage can survive the disappearance of data: information has been archived in an analog way since the time of the first graffiti in the Upper Palaeolithic and today we have way more efficient methods.

This is just to say that data are not inevitable and also the realization of a *digital society*, where control over the information is in the hands of a limited number of subjects, is not inevitable.

11.6 Computers of the Future: The Atoms Strike Back

Overall, there is no doubt that digital technology has revolutionized our society by promoting a progressive loss of weight, through a widely diffused dematerialization process. Let us take a look at the way we enjoy music, for example. In the eighteenth century, a wealthy person who wanted to listen to music, as he pleased, had to host a quartet in his/her house. At the end of the nineteenth century, thanks to Edison's invention, it was sufficient to own a phonograph with waxed cardboard cylinders: undoubtedly a device much lighter than the quartet. Subsequent inventions have all proceeded in this dematerialization path to the point that, currently, digital music is downloaded as electromagnetic radiation into few-gram portable devices. As it is nowadays common experience, dedicated digital tools have also dematerialized newspapers, films, books, paintings, sculptures and, of course, money.

This progressive loss of weight did not come unannounced. One of the famous *Six Memos for the Next Millennium* [5] by Italo Calvino (1923–1985) is devoted to lightness. This lecture was scheduled to be presented for the Charles Eliot Norton Lectures at Harvard, in the fall of 1985, but never delivered because Calvino died unexpectedly on 19 September of that year.

In this first lecture, devoted to the opposition between lightness and weight, the author upholds the virtues of lightness. "*This does not mean that I consider the virtues of weight any less compelling, but simply that I have more to say about lightness. (…) my working method has more often than not involved the subtraction of weight. I have tried to remove weight, sometimes from people, sometimes from heavenly bodies, sometimes from cities; above all I have tried to remove weight from the structure of stories and from language*".

The work program announced by Calvino has been taken quite seriously by those who, in the following years, have designed and distributed the digital tools that today we so widely use, with the unexpected result of flooding us with an enormous amount of digital data.

The dematerialization process we witnessed in the last forty years was aimed at removing information from atoms and transferring it into bits. In my humble opinion, it is coming the time in which things will be turned around.

Remarkably, few initial steps in this direction have already been taken by the pioneers of 3D printing. In a 2008 book entitled *Fab: The Coming Revolution on Your Desktop—from Personal Computers to Personal Fabrication* [6], Neil Gershenfeld, the director of the MIT Center for Bits and Atoms, discussed the potentiality of the fabricators, the new generations of nerds interested in making things: real things, not virtual ones. A clear sign that the atoms revenge has begun.

What is going to be the future of digital technology, then? To answer this question, it comes in our help, again, Italo Calvino. In the same lecture, Calvino speaks about the important legacy of the roman poet Lucretius (c.99–c.55 BC) and cites a beautiful verse from his description of the material world: *the little motes of dust swirling in a shaft of sunlight in a dark room* (II.114–124). What a beautiful description of something that is light without being empty. What an unlimited source of information there is in this dust swirling, impossible to contain in a finite digital representation and, still, so fundamental in our description of reality.

Fluctuation is the word that comes to my mind when I listen to such a description. Fluctuation and noise: exactly what we wanted to get rid of at the beginning of the digitalization process. If it is true that the future belongs to those who can imagine it, we need more noise and fluctuation to get a grasp at the un-simplified reality we live in. No digital representation will ever be able to embrace it. Analog technology will eventually take over digital technology, and for this to happen, we will need a new generation of scientists.

References

1. Computing study refutes famous claim that 'information is physical' https://phys.org/news/2016-07-refutes-famous-physical.html
2. *Proceedings of the 3rd International Symposium on Foundations of Quantum Mechanics*, Tokyo (1989), pp.354–368. https://philpapers.org/archive/WHEIPQ.pdf
3. H. Hosni, A. Vulpiani, Forecasting in light of big data. Philos. Technol. **31**, 557–569 (2018)
4. L. Gammaitoni, Why it is difficult to find something, if you do not know what you are looking for (2021), https://medium.com/geekculture/why-it-is-difficult-to-find-something-if-you-do-not-know-what-you-are-looking-for-1e7ca2f69da5?sk=d614318ca56c9822f9591f626d194944
5. I. Calvino, *Six Memos for the Next Millennium.* (Harvard University Press, 1988)
6. N. Gershenfeld, *Fab: The Coming Revolution on Your Desktop - from Personal Computers to Personal Fabrication.* (Basic Books, 2007)

will eventually take over digital technology, and for this to happen, we will need a new generation of scientists.

References

1. Computing study refutes famous claim that 'information is physical' https://phys.org/news/2016-07-refutes-famous-physical.html
2. *Proceedings of the 3rd International Symposium on Foundations of Quantum Mechanics*, Tokyo (1989), pp.354–368. https://philpapers.org/archive/WHEIPQ.pdf
3. H. Hosni, A. Vulpiani, Forecasting in light of big data. Philos. Technol. **31**, 557–569 (2018)
4. L. Gammaitoni, Why it is difficult to find something, if you do not know what you are looking for (2021), https://medium.com/geekculture/why-it-is-difficult-to-find-something-if-you-do-not-know-what-you-are-looking-for-1e7ca2f69da5?sk=d614318ca56c9822f9591f626d194944
5. I. Calvino, *Six Memos for the Next Millennium.* (Harvard University Press, 1988)
6. N. Gershenfeld, *Fab: The Coming Revolution on Your Desktop - from Personal Computers to Personal Fabrication.* (Basic Books, 2007)